Business Innovation in Asia

Industrial competition with rising economies, new regional investment from the West, and trade pacts among competitors threaten Japan's long postwar prominence. Global market dynamics and regional competition prompted the shift from offshore factories to local networks in the last decade. Similar forces are driving the recent formation of regional Nikkei – Japan-affiliated – nodes in major industrial clusters in Asia.

The central concept of this volume, "knowledge networks," refers to interactive linkages around nodes of tacit and codified knowledge embedded in global value chains. Through survey evidence and interviews at firms and factories, this book reveals the problems facing knowledge transfer, such as persisting difficulties in communication, technology transfer, indigenous learning in regional nodes of Nikkei value chains and the persistence of earlier patterns of hierarchical coordination in information flows despite the shift towards more horizontal network organization. However, a comparison of Nikkei knowledge networks in China, South Korea, and Thailand reveals the possibilities of an interactive learning community in cross-border investment. If Japan can meet the challenge of tapping Asia's offshore resources for innovation, it will pose a formidable global challenge to Western competitors.

This book will be of interest to academics, postgraduate students and professionals working in the field of Asian business, innovation, globalization, and network theory.

Dennis McNamara is the Park Professor of Sociology and Korean Studies, and Special Assistant to the University President for China Affairs.

Routledge Contemporary Asia Series

1 **Taiwan and Post-Communist Europe**
Shopping for allies
Czeslaw Tubilewicz

2 **The Asia-Europe Meeting**
The theory and practice of interregionalism
Alfredo C. Robles, Jr

3 **Islamic Legitimacy in a Plural Asia**
Edited by Anthony Reid and Michael Gilsenan

4 **Asian-European Relations**
Building blocks for global governance?
Edited by Jürgen Rüland, Gunter Schubert, Günter Schucher and Cornelia Storz

5 **Taiwan's Environmental Struggle**
Toward a green silicon island
Jack F. Williams and Ch'ang-yi David Chang

6 **Taiwan's Relations with Mainland China**
A tail wagging two dogs
Su Chi

7 **The Politics of Civic Space in Asia**
Building urban communities
Edited by Amrita Daniere and Mike Douglass

8 **Trade and Contemporary Society Along the Silk Road**
An ethno-history of Ladakh
Jacqueline Fewkes

9 **Lessons from the Asian Financial Crisis**
Edited by Richard Carney

10 **Kim Jong Il's Leadership of North Korea**
Jae-Cheon Lim

11 **Education as a Political Tool in Asia**
Edited by Marie Lall and Edward Vickers

12 **Human Genetic Biobanks in Asia**
Politics of trust and scientific advancement
Edited by Margaret Sleeboom-Faulkner

13 **East Asian Regionalism from a Legal Perspective**
Current features and a vision for the future
Edited by Tamio Nakamura

14 **Dissent and Cultural Resistance in Asia's Cities**
Edited by Melissa Butcher and Selvaraj Velayutham

15 **Preventing Corruption in Asia**
Institutional design and policy capacity
Edited by Ting Gong and Stephen Ma

16 **Expansion of Trade and FDI in Asia**
Strategic and policy challenges
Edited by Julien Chaisse and Philippe Gugler

17 **Business Innovation in Asia**
Knowledge and technology networks from Japan
Dennis McNamara

Business Innovation in Asia

Knowledge and technology networks from Japan

Dennis McNamara

Routledge
Taylor & Francis Group
LONDON AND NEW YORK

First published 2009
by Routledge
2 Park Square, Milton Park, Abingdon, Oxon OX14 4RN

Simultaneously published in the USA and Canada
by Routledge
270 Madison Ave, New York, NY 10016

Routledge is an imprint of the Taylor & Francis Group, an informa business

Typeset in Times New Roman by Pindar NZ, Auckland, New Zealand

British Library Cataloguing in Publication Data
A catalogue record for this book is available from the British Library

Library of Congress Cataloging-in-Publication Data
Business innovation in Asia: knowledge and technology networks from Japan / Dennis McNamara
p. cm. — (Routledge contemporary Asia series)
1. Technological innovations—Economic aspects—Japan. 2. Business networks—Japan. I. McNamara, Dennis L.
HC465.T4B87 2009
338'.0640952—dc22 2009002000

ISBN10: 0-415-49935-6 (hbk)
ISBN10: 0-203-87426-9 (ebk)

ISBN13: 978-0-415-49935-4 (hbk)
ISBN13: 978-0-203-87426-4 (ebk)

Contents

Tables

Preface

Urgent issues of innovation at firms, knowledge networks in societies, and community in the region have come to the fore in East Asia. Prospects for a new knowledge nexus in East Asia have stimulated a wide range of discussions. Intriguing questions of security, national development, and global trade and investment have fueled a vigorous debate. Some look to knowledge as power and turn quickly to issues of national security. Others see opportunity for moving up the development ladder with cross-border flows of capital, expertise and knowledge workers. Many in Japan struggle with the nation's future in the region, and the new technological competition from her neighbors. Hopes for an East Asian Community have spurred still others to plans of knowledge networks binding a region beyond either markets or militaries. Similar questions have drawn me across a decade of study in bringing this research project to print.

I began at the intersection of national development and cross-border production chains to learn more of the interaction of local and global dynamics. Technology transfer in manufacturing chains was my focus. But as I witnessed the rise of the "New Asia," regionalism captured my attention. Discussion at the initial East Asian Network of academics (NEAT) under the ASEAN + 3 umbrella drew me deeper into issues of community formation. Colleagues at the annual meetings of the APEC Academic Centers looked more concretely to policy and practice in investment. A rapidly expanding set of bilateral trade pacts gave new urgency to the debates, and concrete form to cooperation and even coordination across borders.

While others lauded regional solidarity, economic development, or the growing intensity of intra-regional trade and investment, I was finding a far more interesting transition. I was looking for technology transfer but finding much more. A variety of factors such as the outsourcing of research and development by Western firms, improved communication networks, and globalization and structural reform among Japanese firms, all contributed to new knowledge flows among Japanese firms in traditional manufacturing industries. Moving among firms, industry associations, and trade organizations, I was witnessing the beginnings of a remarkable shift in Asia from simply offshore manufacturing chains to knowledge flows.

Innovation brought these disparate themes into a focus. A chronology of Nikkei networks in three industrial sectors profiles the major changes. Portraits of leading Japanese firms and their leadership give texture to the story, just as innovation systems in the three nations of interest provide context. Issues of complementarity, cooperation and competition, and even coordination in the new preferential trade agreements, all provide markers for contrasting knowledge networks in the three nations. I look to major Asian nations hosting Japan's foreign investment to assess interactive learning for firms across borders. Home firms in electronics, autos, and textiles give texture to the picture abroad. Why do some host nations manage to anchor productive activity, and still others innovation?

A profile of process and prospect emerges for a shift from assembly abroad to knowledge exchange in Japan's offshore manufacturing networks. I open the discussion, pose the questions, and chronicle a major transition in the region. Precedent suggests prospect, and may contribute to constructive policy and practice. Will national security trump regional synergies? Will Japan carve out a new leadership role in tandem with rising powers such as China and South Korea? How would the West meet the challenge of a new knowledge nexus in East Asia?

* * *

It is difficult to single out people and organizations across a decade of study that have contributed to my work. The Fulbright-Hays Program of the US Department of Education supported research in Seoul and Tokyo. The Korea Foundation supported meetings of the Georgetown University Conference on Korean Society, a forum for drawing out many of these ideas. The East Asian Institute of Thammasat University in Bangkok brought me into the early formation of the East Asian Community, and organized seminars around the topic. The Institute of Industrial Economics at the Chinese Academy of Social Sciences in Beijing helped with research support and a group of colleagues who pressed me forward on issues of regionalism and development.

Equally significant in a project of this length and breadth are the university colleagues who help to sustain one's interest and imagination. Faculty at the Graduate School of International Studies at Sogang University in Seoul, and at Sophia and Waseda in Tokyo certainly played that role. Students and faculty alike at Renmin University and Fudan University in China, and Thammasat University in Bangkok brought still new challenges and insights to my work. And finally, a word of thanks to my colleagues at Georgetown University who joined various seminars, conferences, and other fora, to bring insight and wisdom to the project.

Abbreviations

ACFIF	Asian Chemical Fiber Industries Federation
ADB	Asian Development Bank
AFTA	ASEAN Free Trade Agreement
AICO	ASEAN Industrial Cooperation
AOTS	Association for Overseas Technical Scholarship
APEC	Asia–Pacific Economic Cooperation
Art.	Article
ASEAN	Association for Southeast Asian Nations
ASPAC-TCIF	Asia Pacific Textile and Clothing Industries Forum
ATB	Asian textile business
BOI	Board of Investment, Thailand
BOJ	Bank of Japan
BP	*Bangkok Post*, Thailand
CACFI	Conference on Asian Chemical Fiber Industries
CCPIT	China Council for Promotion of International Trade
CD	*China Daily*
EAEP	*East Asian Economic Perspective*
EEI	Thailand Electrical and Electronics Institute
EOI	export-oriented industrialization
EPA	Economic Partnership Agreement
ERIA	Economic Research Institute for ASEAN and East Asia
ESCAP	UN Economic and Social Committee for Asia and the Pacific
FDI	foreign direct investment
FKI	Federation of Korean Industries
FTA	Free Trade Agreement
FTI	Federation of Thai Industries
GERD	gross domestic expenditure on R&D
GM	General Motors
GMB	global market briefings
GTC	general trading companies
GTIS	Global Trade Information Service
GVCs	global value chains
ICT	information and communication technology

IDE	Institute of Developing Economies, Japan
IFDI	inward foreign direct investment
IMF	International Monetary Fund
IMV	international multi-purpose vehicle
INEF	Institute for Development and Peace, University of Duisburg
INT	Interview
IPR	intellectual property rights
IRI	Industrial Research Institute Company
ISI	import-substitution industrialization
ISO	International Organization for Standardization
ITGLWF	International Textile, Garment, and Leather Workers' Federation
JAMA	Japan Automobile Manufacturers' Association
JAPIT	Japan Association for Promotion of International Trade
JBIC	Japan Bank for International Cooperation
JCEA	Japan–China Economic Association
JCFA	Japan Chemical Fibers Association
JCHE	Japan Chemical and Heavy Industries News Agency
JEITA	Japan Electronics and Information Technology Industries Association
JEMA	Japan Electrical Manufacturers' Association
JETRO	Japan External Trade Organization
JICA	Japan International Cooperation Agency
JRI	Japan Research Institute
JSBRI	Japan Small Business Research Institute
JTECS	Japan–Thailand Economic Cooperation Society
JTEPA	Japan–Thailand Economic Partnership Agreement
JTEPO	Japan–Thailand Economic Partnership Agreement Office
KCFA	Korea Chemical Fibers Association
Keidanren	Japan Business Federation
KIEP	Korea Institute for International Economic Policy
KNSO	Korea National Statistics Office
KOFOTI	Korea Federation of Textile Industries
KT	*Korea Times*
LCD	liquid crystal display
LDI	LCD (liquid crystal display) Driver IC (integrated circuit)
MERI	Mitsubishi Economic Research Institute, Japan
METI	Ministry of Energy, Trade and Industry, Japan
MEW	Matsushita Electric Works
MEXT	Japan, Ministry of Education, Culture, Sports, Science and Technology
MITI	Japan's Ministry of International Trade and Industry, predecessor to METI
MKE	South Korea's Ministry of the Knowledge Economy
MNE	multinational enterprise

MOCIE	Ministry of Commerce, Industry and Energy, Korea
MOST-C	Ministry of Science and Technology, China
MOST-K	Ministry of Science and Technology, Korea
NBSC	National Bureau of Statistics China
NEAT	Network of East Asian Think-tanks
NIEs	newly industrializing economies (Taiwan, Hong Kong, Singapore, and South Korea)
NSOT	National Statistics Office, Thailand
NSTDA	National Science and Technology Development Agency, Thailand
OBM	own brand manufacture
ODA	overseas development aid
OECD	Organization for Economic Cooperation and Development, Paris
OEM	original equipment manufacture
PDP	plasma display panel
PTA	purified terephthalic acid
RCAPS	Ritsumeikan Center for Asia Pacific Studies
RIETI	Research Institute of Energy, Technology, and Industry (Japan)
SAIC	Shanghai Automotive Industry Corporation
SCB	Statistics Bureau, Japan Ministry of Internal Affairs and Communications
SFERTC	Shanghai Foreign Economic Relations and Trade Commission, China
SFIC	Shanghai Foreign Investment Commission, China
SMEA	Small and Medium Enterprise Agency, Japan
SMRJ	Organization for Small and Medium Enterprises and Regional Innovation, Japan
SWAK	Spinners and Weavers Association of Korea
TAB	tape automated bonding
TAI	Thailand Automotive Institute
TAIA	Thai Automotive Industry Association
TAPMA	Thai Autoparts Manufacturers' Association
THTI	Thailand Textile Institute
TKS	Tōyō Keizai Shimbunsha
TMC	Toyota Motor Corporation
TN	*The Nation*, Thailand
TNI	Thai-Nichi Institute of Technology
TPA	terephthalic acid
TPA	Thailand–Japan Technology Promotion Association
TRIPS	Agreement on Trade Related Aspects of Intellectual Property Rights
TSMA	Thai Synthetics Fiber Manufacturers' Association
TSST	Toshiba Samsung Storage Technology Corporation
TWARO	Textile Workers Asian Regional Organization

UMTS	Universal Mobile Telecommunications System
UNCTAD	United Nations Conference on Trade and Development
UNIDO	United Nations Industrial Development Organization
WIPO	World Intellectual Property Organization
WTO	World Trade Organization

1 Business, knowledge, and networks

A global competition for knowledge is driving change in Asian markets and societies. Just as finance capital once ruled in industrial society, Daniel Bell recognized the leading role of information and knowledge in the post-industrial society (1973; Kumar 1995). The forms of *knowledge* may be familiar, but the linking of knowledge in new communication channels is indeed revolutionary. So too is the global reorganization of manufacture where knowledge is controlled and shared largely within global value chains of production and marketing. Fruin wrote of a novel and very fluid type of "network society," paralleling the electronic network of multiple nodes without a permanent center and no longer bounded by geographical setting. Networks might be seen as "sets of independent actors who cooperate frequently for mutual advantage and create a community of practice (1998: 4)." Knowledge surging in global streams across national borders makes possible new types of communities of practice.

Information and communication systems now make it possible to coordinate production across multiple sites, enabling firms to reorganize into global value chains (GVCs) beyond the earlier model of vertically integrated production within national borders. Networks permit specialization but also flexibility and responsiveness to changing consumer demands. An additional dynamic of clustering into major production nodes has promoted the reorganization of now hyper-mobile international capital. Multinational enterprises will invest in production at major sites abroad with attractive infrastructure and labor costs, leading to a concentration of suppliers for some regions, and isolation for others. A new competition for knowledge resources forces difficult choices of inclusion or exclusion, cohesion or fragmentation in the globalization process. Successful strategies for inclusion among developing nations in the "New Asia" merit closer scrutiny.

Technology has played a central role in the rise of East Asia to prominence in global production chains. Effective acquisition and application of ever more sophisticated technology distinguished the postwar recovery of Japan, and subsequently the trajectory of the "Newly Industrializing Economies" (NIEs) of Taiwan, Hong Kong, Singapore, and South Korea. Access to foreign technology helped to fuel China's economic trajectory, and spur development of local institutions of innovation. Yet other regional economies such as Thailand have seen far less growth in local industry despite decades of foreign investment. On the demand side, growth in urban consumer markets now complements export growth, and has begun to affect inward foreign direct investment (FDI) across the region. One recent change in Japanese offshore manufacturing in Asia is a shift of horizontal networks abroad, so that a full line of manufacturing is possible on site. An earlier vertical pattern of manufacturing segmented between higher value-added aspects in Japan, and lower value-added assembly abroad (METI 2005). Growing local market demand and the easing of trade restrictions help to drive the change. Market integration in a "seamless economic zone" has brought manufacturing closer to markets (METI 2007d). Liberalization and preferential trade agreements make possible close to tariff-free supply of raw materials and intermediate goods across borders within the Association for Southeast Asian Nations (ASEAN). This has intensified the sharing of production across national borders in dense networks of manufacture and distribution, a further feature of this "New Asia."

Lower labor and energy costs, extensive human resources, and relative geographical proximity of countries within the region encouraged extension abroad of manufacture, first with textiles and clothing, and then with automobiles and electronics. Major clusters of production have emerged on the outskirts of Seoul, Shanghai, and Bangkok. Growth in foreign and local demand has increased both the scale and technological levels of production, prompting deeper, more extensive flows of knowledge across the region, a final feature of the "New Asia" and the subject of this study. We find evidence of the latter not only in international comparisons of national innovation systems, but also in product development such as the jump from electrical to electronic products, fuel-efficient autos, and industrial textiles. Profile and prospect of regional knowledge networks come into view at the intersection of local systems of innovation with GVCs in electronics, autos, and fashions.

Technology

Technos or "technique" includes organizational strategies and machines. One scholar defined the transfer of technology as "a learning process where technological knowledge is continually accumulated into human resources engaged in productive activities." Shiowattana enumerated human resources such as acquisitive, operative, adaptive, and innovative capacity (1990: 112). Transfer of management and production techniques with relevance beyond the single firm or industry can benefit local society. For instance, Japanese management strategies often include job rotation rather than fixed duties and specialization. A community of multi-process or multi-function workers committed to a firm is the goal. A related feature is the principle of "floor-level" (*genba*) or "hands-on" experience which brings managers, technicians, and assembly line workers to the production line to learn of the various functions in assembly.

Certainly this can promote community among workers, marketing and engineering divisions, but it is likewise a critical channel for skill transfer in the local community. Monden wrote of just-in-time production and quality control in *Toyota Production System* (1998). The former refers to central coordination of supply and production schedules to reduce costs of inventory deficiencies and line shutdowns. Delivery depends on information and transportation systems in the local society that seldom match the efficiency of Japan. Yet transfer of coordination techniques and scheduling systems can benefit local suppliers. Equally important is autonomous surveillance to prevent defects whether by inspectors or by regular staff as an internalized function of workers along the production line. Koike argued that the "diagnostic ability to detect causes of manufacturing defects" was the key transfer for Japanese firms moving abroad (Koike 1997: 46).

Acquisition of new machinery and production materials can benefit the local firm and industry, especially if supply contracts promote local learning and local calibration of the machinery. Sub-contracting within a network of long-term suppliers is a further feature of Japanese industrial organization. Vertically integrated production in the US auto industry has often been contrasted with the preference of Japanese assemblers for a set of affiliated first-tier suppliers, supported by a skein of smaller sub-contractors. But how do you select partners with whom to share proprietary technology? Recent efforts to reduce time and development costs by joint-development of new technologies with parts suppliers have given the issue even greater salience. Michael Smitka argued that a "conscious reliance on trust" as a tool for governance in contracting ties was among the most important innovations in Japan's auto industry. Ties of trust between assembler and supplier permitted "extension of innovations from intra-firm to inter-firm transactions (1991: 4, 10)." Transfer depends on scale of ownership and extent of control. Ownership appears a major factor in promoting the trust necessary for the transfer of technology among Japanese firms moving abroad. Majority equity participation is often correlated with transfer of critical technologies (Tho 1992; Tsuneishi 1990). Beyond capital investment, Perrucci cited a further condition promoting transfer: a "network of supportive social relationships between groups with economic and non-economic interests (1994: 18)."

Japan's commitment to technology as a form of national strength and security gave direction to remarkably cohesive goals of development across three decades from the 1950s. Japan's long-cherished ideal of the world's factory with closely held technology began to fade with the Plaza Accords of 1985 and resulting appreciation of the yen. Comparative cost advantages drove Japanese firms large and small abroad to their Asian neighbors. The next decade saw the integration of these offshore factories into vertical networks of production controlled from headquarters in Japan. Some have faulted the networks for retaining the best technology in Japan to maintain a hierarchy of knowledge and value-added production. Others counter that Japanese firms excelled at basic training programs, and indeed that the Japanese government's investment in infrastructure promoted development of "supporting industries" within ASEAN and China. The term denotes a "mountain-like structure" of mainly small and medium enterprises (SMEs) supporting larger firms in manufacturing through specialized production but also product development and even research.

As the region roared back from the Financial Crisis of 1997, the system has evolved into nodes evident in horizontal rather than vertical integration of regional production. Efforts at technology or "knowledge transfer" appear all the more important today with the third stage in the evolution among Nikkei offshore subsidiaries. Government reports and surveys document the difficulty of knowledge transfer in the trajectory from offshore factory to integrated network, and now to network nodes. Precedents of "cooperative learning" and "supporting industries" in Japan provide a Nikkei model for knowledge transfer abroad. I track the policy and progress of the model abroad in Asia through survey data at the Ministry of Energy, Trade and Industry, Japan (METI), the Japan External Trade Organization (JETRO) and the Japan Bank for International Cooperation (JBIC). Counterpart organizations in host countries balance the picture. Technology offices in Japan, China, South Korea and Thailand keep track of knowledge promotion and knowledge profiles at home and in the region, drawing on data from the World Intellectual Property Organization (WIPO). Reports at home and host country industry associations provide further extensive detail on "localization" of management and procurement, as well as on research policy and investment. Government reports on support for the internationalization of SMEs offer a detailed picture of Japanese efforts to foster "supporting industries" abroad similar to the cooperative model in the home islands. The materials shed new light on the prospect and problems of Japan's knowledge flows in China, South Korea, and Thailand. Improving Nikkei knowledge networks in the New Asia is Japan's next challenge.

Knowledge transfer

Wide attention to innovation and corporate management of knowledge, and to its creation and diffusion among state leaders, reflects the growing consensus that knowledge is indeed the key to productivity. Sources of economic value are shifting towards "the creation, acquisition and use of intellectual assets (METI 2004:16)." Steinmueller argued that knowledge supports improvements in productivity and organization, and fosters new industries (2002). Technology transfer in cross-border manufacturing depends on modes of transmission, but more fundamentally, on the type of knowledge transferred. For instance, we might describe what we "know" as an objective reality, or as a subjective belief, dependent on a particular set of experiences. Nonaka and colleagues argued the latter "constructivist" approach in defining knowledge as "justified true belief," conditioned by the subject's point of view, their sensibility, and the character of their experiences. He gave priority to the context that "harbors meaning," a common time and space in which knowledge is created and shared (2001b: 13–14).

Subjective construction and a common context help to define a process for learning among firms. We can describe the learning process simply as the acquisition of knowledge beyond the boundaries of the individual firm (Wijnhoven 2006). Polanyi's enduring insight that "we know more than we can tell," perhaps best conveys the necessity of mutual learning for home and host firms as we stretch manufacturing chains across nations, cultures, and firms (1967). Radosevic cited the utility of the contrast between *codified* or explicit knowledge with *tacit* knowledge in technology transfer efforts (1999). *Codified* knowledge refers to "know-what (facts) and know-why (scientific principles)," whereas *tacit* knowledge refers to "know-how (personal skills) and know who (how and where to get specialized knowledge)." Teaching "what" or objective facts, is far easier than communicating know-how, and learning of the former proves far easier than acquisition of the latter (Asheim and Isaksen 2000: 171).

Knowledge flows along two dimensions within the chains. A vertical flow of information develops between the multinationals and the local suppliers. Explicit knowledge of organization and technology moves along this axis in the form of specifications and manuals, and often through seminars or even training sessions at the headquarters of the multinationals. Transfer of tacit knowledge is more difficult along this vertical axis, given differences in culture and experience, and limited opportunities for the shared space and time of communication. A horizontal stream of knowledge helps integrate a cooperating group of supporting industries within the target nation. Continuities of location, ethnicity, schooling, or religion, often promote a cultural basis for transferring tacit knowledge as well – a critical feature of social systems of production. The latter refers to the distinctive institutional arrangements of corporations, trade associations, and state offices in any one nation. An emphasis on a unique history or "path-dependent logic" in the evolution of a nation's system of production provides a balance to premises of homogenization across borders suggested in some studies of globalization (Hollingsworth and Boyer 1997b). Socially embedded norms and rules deeply affect the knowledge flows in the anchoring of manufacturing investment from abroad.

Institutions such as foreign and local business associations and labor organizations, as well as industry offices in government can promote the deepening of industrialization in the shift from clusters to what we might term "community (Brown and McNaughton 2002; McDonald and Vertova 2002)." The shift will demand resources and take time, but remains critical for local knowledge and technology transfer. External linkages often appear the priority in cluster promotion as government offices scramble to attract foreign investors and their technology. But weak internal ties linking business and social networks among local and foreign SMEs severely curtail the "backward linkages" or spread effects of learning from foreign investment and technology. Specialists at the United Nations (UN) explained that "the direct effect of linkages on domestic suppliers is generally a rise in their output and employment. Linkages can also transmit knowledge and skills between the linked firms (UNCTAD 2001: 15)." Governments compete to anchor mobile international capital with spillovers of technology and organizational capability. Narula and Dunning distinguished between conventional FDI exploiting natural resources or penetrating markets abroad, from a newer type of FDI seeking "created-assets" in host countries, such as the nation's knowledge resources (2000). The question today in Asia is not so much separate FDI investments in knowledge-intensive firms and sectors, but rather the upgrading of production networks with more intensive knowledge exchange.

Networks

Nodes or hubs within knowledge networks of value chains provide a distinctive setting for knowledge transfer. Market, hierarchy, and network represent three distinct forms of organization. Markets offer an impersonal setting for resource allocation with goods exchanged according to pricing mechanisms. Mass production has prospered in hierarchical firms with clear lines of authority, departmental responsibilities, and tracking procedures. Powell explained that networks excel in flexible, customized production, operating without discrete exchanges or administrative fiat, but rather "through networks of individuals engaged in reciprocal, preferential, mutually supportive actions (1990: 103)." Markets permit exchange of codified knowledge such as patents and licenses. Hierarchical organization within a firm prospers with exchange of both codified and tacit knowledge. But if networks can establish necessary bonds of trust, collaboration in both codified and tacit knowledge can be fostered without the cumbersome bureaucratic hierarchy necessary for intra-firm cooperation. Thus much of the network literature in sociology looks to mediating structures such as industry associations, guilds, or ethnic ties of local area or national context to support "social capital." A parallel can be drawn with Japan's rich tradition of dense networks linking larger and smaller firms, center and periphery, and government and business.

Japan's wider social system has been termed a *network* that joins formal network organizations with informal societal networks (Kumon 1992). A nation-wide societal network, or "Nihon," with institutionalized channels for effective exchange of information can be instrumental in learning and innovation. Looking to networks and knowledge, Morris-Suzuki gave priority to the social context of innovation among SMEs and its contribution to development of manufacturing technology (1994). At the level of firms, James Lincoln looked to technology in the horizontal network (*kyōeikai*) suppliers at the home electronics giant Panasonic[1] (2006). In contrast, Toyota's capital investment in its first-tier suppliers such as Denso results in a vertical network of component supply for assembly. Denso in turn organizes "cooperatives (*kyōryokukai*)" of first- and second-tier suppliers which specialize in production and product development. Suppliers or "sub-contractors" in both sectors often play a role in the design and planning stages well before actual manufacture.

Williamson highlighted "credible commitments" beyond contract as a means of reducing the costs of transactions to the embedding of economic exchange in social institutions. Putnam later extended the focus to "social capital" arising from civic engagement, as a means to trust in exchange relationships (1993). Substantive norms and dense networks of generalized reciprocity appear particularly strong in Japan, where Ron Dore and others had earlier alerted us to the strength of social bonds supporting cooperative forms of competition in the market (1983). Most of this literature remains limited to case studies of single nations, sorting out the institutional and wider cultural bases for solidarity in the market. Technology transfer within a Nikkei network demands a comparative analysis of institutional patterns fostering trust from both the sending and receiving sides.

Recent scholarship on networks has brought new insight into the structure of social capital, its benefits and limits. Portes wrote that "social capital stands for the ability of actors to secure benefits by virtue of membership in social networks or other social structures (1998: 6)." Haggard and colleagues suggested three defining features, such as "dense networks, norms of trust, and generalized reciprocity (1997)." Functions add further detail to the concept, and can include "fostering trust, fine-grained information transfer, and joint problem-solving." Others suggest benefits of influence and solidarity (Uzzi 1996). One might ask how firms can balance embeddedness with autonomy, particularly given the asymmetric relationship between Japan and her Asian partners. Moreover, one result of the debate over "closure versus network holes" has been the recognition that too much social capital might lead to overspecialization and exclusion from other productive networks.

Group closure may deprive the group of access to new sources of information. Exclusive sub-contracting ties in Japan's auto industry, for instance, permit SME participation in basic product development (*seihin katihatsu*) with major firms. On the other hand, too little social capital would preclude participation in those networks in the first place. Granovetter argued that "weak ties," i.e., less structure, less closure, and less redundant links provide access to more distant networks with information unavailable within one's more familiar circles. One might expect, for instance, strong corporate ties with offshore, wholly owned or majority-owned subsidiaries, but weaker ties within sub-contracting networks abroad. Finally, we need to gauge as well the diachronic dimension of network development, or the "coupling and decoupling" of ties as firms and markets change (Granovetter 1995). Firms shed confining ties and forge new links as their need for new inputs and wider market opportunities develop.

Coordination of cross-border networks involves new forms of trust beyond contract. Hollingsworth and colleagues defined *coordination* as the "totality of institutional arrangements – including rules and rule-making agents – that regulate transactions inside and across the boundaries of an economic system." Problems common in developing countries such as uncertainty of legal structures, ambiguity or discontinuity in government policies on trade and investment, would force higher transaction costs. More tangible and conventional forms of coordination to reduce costs include markets and corporate hierarchies, where prices or corporate hierarchies govern transactions. Hollingsworth and colleagues found that informal networks, on the other hand, remain relatively undefined due to "particularistic, mutually obligated, and legally non-enforceable relationships (1994: 4)." Far less structured than vertically integrated firms, and yet with more intense interactions and closer interdependence than impersonal market transactions, networks depend upon sources of social capital embedded in the society.

One key to network interactions is compatibility between social systems of production in investing and host, particularly of foreign firms with local social and political networks. Local intermediaries play a role in Nikkei networks, paralleling the roles of state and business association. Foreign investment includes wholly or majority owned subsidiaries, subsidiaries with investment and technology ties, and licensing of technology. Granovetter identified the density of associational life as a key component of successful network ties, particularly the strength of cross-cutting ties across polity and economy. But how does one gauge the strength, or define the institutional patterns of such ties? Doner highlighted Japan's distinctive style of economic cooperation in Asia with a "densely institutionalized network for information exchange and policy coordination" between government and private sector (1997: 203–4).

Biggart and Guillén studied social organization in the auto industry to assess links to the global economy. State policy and market practice, and the scale of corporate organization, dominated their analysis, particularly in South Korea's "institutional logic" where the state favored large conglomerates at the expense of small and medium-sized firms (1999). Unger highlighted the role of social capital among ethnic Chinese in Thailand as a key to business–state cooperation in promotion of industries, and induction of foreign investment (1998). Wang's study of transitions in China's political economy emphasized the role of social capital in foreign investment in the absence of credible and consistent structures of law (2001).

Networks permit cooperation within flexible, permeable boundaries. Nodes or hubs represent a critical point of interchange. Networks vary greatly among societies and industries, with Japanese networks prominent as an enduring form of societal organization permitting the sharing of knowledge. How do these networks operate in manufacturing investment abroad, and what does this tell us of Japan and of the region? It is not the Nikkei networks alone that draw my attention, but rather the interaction of Nikkei and local innovation systems in East Asia. I look to knowledge networks across rather within firms, and on acquisition of "external knowledge." *Knowledge networks* can be defined as interactive linkages around nodes of tacit and codified knowledge embedded in GVCs. Such networks can be distinguished by the process or format of information exchange, the organization of the networks within firms, and by target market or product. The networks provide a window on the interaction of global investment and local social systems of production, or more broadly, linking of a space of flows with a space of place.

Global flows

We have witnessed a remarkable shift from the extension of firms abroad to the integration of global networks among those same firms across the past four decades. Peter Dicken wrote of a global system of coordinated production today at different sites across the world based on local endowments of labor, resources, and market in line with comparative advantage. Globalization no longer means simply trade or investment across national boundaries, but rather "functional integration of internationally dispersed activities (2003: 5)." In a similar vein, Leslie Sklair contrasted the site-bound production of earlier multinationals, with regionally and globally integrated production that links transnational corporate networks without regard to national boundaries (2001: 143). Dicken and Sklair, however, differed profoundly on the role of borders. The former emphasized the significance of conflict and collaboration between state and multinational enterprises (MNEs).

I extend Dicken's analysis to state-to-state collaboration in regionalization, as well as to ties between state and firm. Sklair dismissed "state-centric" interpretations of investment by MNEs guided in part by national interests, in favor of a focus on the transnational capitalist class interested primarily in global profit and market share. But even Sklair highlighted efforts of MNEs to harmonize their aims with local goals of development whenever possible: i.e., to "think global and act local."

MNEs have refocused their energies on shareholders at the expense of constraining national interests. Yet a state role and interest remains prominent today in Japan's techno-nationalism, and indeed in the investment strategies and innovation systems of other Asian nations. The Japan International Cooperation Agency (JICA), the Finance Ministry, and the Ministry of Economy, Trade and Industry (METI) continue to play major roles in economic diplomacy. Policy and practice in the interactions of states within the network, and of semi-official business associations, deserve closer scrutiny. Indeed, the strong state role in the development of Japan and South Korea has been widely recognized. A socialist polity in China keeps close reins on supply and demand, and especially on foreign investment, as it struggles to strike a balance between socialist polity and capitalist market. The relatively weak state role in the earlier development trajectory of Thailand suggests one possible answer to the differential rooting of values chains in China, South Korea, and Southeast Asia. Clarifying state and MNE cooperation and conflict, as well as state-to-state ties will shed light on the relative strengths of global market versus national economic interest in East Asia.

The changing role of states and markets helps to explain the reorientation of firms to global interests. New modes of interaction through dense, overlapping linkages of transportation and communication are driving profound changes in market and society across borders. Michael Storper highlighted a dialectic between "flow economies" and "territorial economies" in contemporary capitalism (1997: 181). Manuel Castells wrote similarly of a space of flows and the space of place. *Flows* refer to "sequences of exchange and interaction" among social actors in physically separate locations. A space of flows denotes "the material organization of time-sharing social practice that work through flows." In contrast, *place* referred to "a locale whose form, function and meaning are self-contained within the boundaries of physical contiguity." Although the logic of flows supersedes earlier dynamics limited by location, the flows today remain rooted in specific locales (1996: 412, 423). The irony of the hyper-mobility of capital and information today among multinational firms is the growing significance of the local.

Localization priorities among offshore subsidiaries suggest the economic significance of place. David Harvey reminded us of the growing significance of local "fragmentations" for social identity and action, in the face of expanding standardization and homogenization across borders. Emerging global unities appear to encourage local distinctiveness. As borders shrink and global capital seeks out production sites, there is a growing appreciation for what makes a place "special and gives it a comparative advantage (1989)." Others cite the dialectical relationship between flows and locales in the shaping and operation of global supply or "value" chains (Henderson 2004). Attention to flows then raises questions of how locations with specific cultural and political contexts of production transform global chains, and at the same time how chains shape local production systems. If, indeed, globalization can lead to fragmentation and isolation, then a dialectic between global and local is possible only if we can assume the locality can gain a place in global flows (Willet 2001). How does a nation anchor a hub or *node* in a global value chain?

Hopkins and Wallerstein singled out production sequences or "commodity chains" as the "warp and woof" of production in the global economy. Chains represent socially determined "networks of labor and production processes whose end result is a finished commodity (1994)." Value chains or streams of production, distribution, and marketing provide a window on globalization (Humphrey and Schmitz 2002: 18). Porter (1990) explained how one company's value chain must be embedded in a larger value system including other companies such as suppliers, distributors, and even buyers. Chains refer to "inter-organizational networks, clustered around one commodity, linking households, enterprises, and states within the world-economy (Gereffi and Korzeniewicz 1994: 2)." Value-added activities performed by interlinked firms might include products, services, and resources (Todeva 2006: 142). Gereffi highlighted critical dimensions such as geographically dispersed locations for manufacture and marketing, a governance structure, and an institutional framework (1995; 2001). I look particularly to two dynamics critical in building knowledge networks: how the chain integrates or isolates, and how the chain adapts to or changes local production systems in the process of foreign direct investment.

Inclusion and adjustment, as well as the broader dimensions, tells us of power among lead firms on the one hand, and growth or development among their networks of offshore subsidiaries and sub-contractors on the other. The term *power* suggests both corporate resources, and the institutional leverage of state and affiliated business associations. The term *development* denotes industrial upgrading associated with transfer of technology and advance to more "technologically sophisticated capital and skill-intensive economic niches." On a broader level, upgrading demands integration into an intra-regional division of labor at levels of firm, industry, and societal systems of production (Gereffi 1999: 52). I look to knowledge networks within global value chains to assess the process of learning in major industrial nodes.

Knowledge is controlled and shared in global markets today largely within GVCs of production and marketing, where larger firms command critical resources of knowledge hierarchies with considerable investment in technology and research. Attention to chains reveals the organization of production and distribution among lead firms and suppliers large and small, and links to institutions of the public and private sector sustaining the flows. Others have looked recently to the transition from production capability to innovation capability in GVCs (Altenburg, *et al.* 2008). I am more concerned with the process of knowledge exchange, whether evident in improved production capability or innovation. My focus is the incremental process of learning, the challenge of imitating and adapting, as well as of creating new knowledge.

Local systems

Some look to global flows to assess power among investors and development among recipients, others to national systems. I look to the intersection of global flows and local systems to explain the process of innovation in Nikkei production chains. Neither global flows nor local practice alone can adequately explain the current challenge to Japanese regional dominance from China and Western multinationals. How has global competition affected patterns of governance in Nikkei networks, and transfer of technology critical for local development? Advantages of Japan's national business system have been widely recognized, whether in patterns of business–state ties, strength of industry associations, labor's identification with the firm, or in corporate strategies such as just-in-time production. Given the global strength and deep regional roots of Nikkei corporations, the transfer of Japanese business practice and standards to Asia might be expected. Japanese business patterns have received extensive scholarly attention, as have Japanese "transplants" to the USA. But little attention has been devoted to the comparative study of Nikkei GVCs across societies, particularly in the Asian region. Only a comparative study can ferret out significant continuity and difference in Nikkei networks among Japan's Asian investment partners. The China challenge, the growing share of East Asia in international trade and investment, and Japan's growing dependence on her offshore Asian production sites all highlight the urgency of a clearer understanding of network power and local development.

Development begins with local context. If capital and information move along a space of flows, the space of place remains embedded in the social fabric and institutions of a particular locale. Even a globalist such as Castells warned that the role of the local can well be lost in the rush to grapple with the global (1996: 413). One issue is power. As a-historical flows of capital and information gradually condition and sometimes supersede the logic of locally embedded institutions with a "logic of global power," that may well trump local interests. What can be said of the relative contribution of endogenous versus exogenous factors in the origins and persistence of local patterns of innovation? Opponents of this global convergence thesis see national business systems as being far more significant than global flows, and remind us of the significance of national origins even in the operation of MNEs (Whitley 1996). A societal focus in the comparative study of business systems permits a more integrated view of business–state ties at home and abroad. A core concept here is the "social system of production" representing the dominant pattern of complex institutions in the economy. These institutions include individual firms and links among firms, sectoral systems of training, finance, labor, and finally state policy. An "institutional logic" becomes apparent in the working of these systems: "institutions are embedded in a culture in which their logics are symbolically grounded, organizationally structured, technically and materially constrained, and politically defended (Hollingsworth and Boyer 1997a)."

"Institutional complementarity" among firms and social systems remains a priority for home and host countries in foreign investment, despite popular assumptions about "harmonization" across borders. Complementarity might be defined more generally as mutual gains from shared efficiencies. Complementarities in local systems provide the continuity of "supporting industries" necessary to attract long-term manufacturing investment. Cohesive government policy and business practice, reliable training and education systems, and rational labor–management relations, for instance, provide the stability necessary to anchor fluid global capital. Interactions of foreign with local business systems evident in networks of suppliers for foreign-invested firms may well lead towards common regional or global standards. But systemic continuities across borders would not preclude the divergence of local comparative advantage, particularly at the level of industrial sectors, or at the level of larger and smaller firms. Indeed, complementarities in local transport and energy supply, in business and trade law, and in investment incentives etc., could alleviate the inefficiencies of conflicting procedures and practices in home and host countries.

Complementarities represent a major advantage in the competition to attract foreign capital. Dunning revised his model of investment incentives to better represent the attraction among developing economies. Among priorities for foreign investors in the original model were power or ownership, location, and the ability to maximize synergies or "internalization." He later added other factors such as a satisfactory legal system, commercial infrastructure, and business culture; transport and communications facilities and human resources; infrastructure and government policy pursued toward direct investment (2000: 131). A similar theme is struck in Japan's JETRO reports which single out the critical role of local "business environments" in attracting Japanese foreign investment (JETRO 2004: 15). The same term appeared prominently in the Joint Study Group Report on trilateral investments among China, Japan, and South Korea. The Group targeted five areas for improvement: "transparency of laws and regulations, protection of intellectual property rights, dispute settlement mechanisms, promotion of investment, and coherence of national and local administration." Market access and low production costs alone do not drive foreign investment. Among Nikkei firms such as Panasonic, Toyota, and Toray, I find further priorities such as credible labor organization and laws, business associations which can serve as counterparts to Japan's associations, and state policy supporting inward FDI. The three issues appear prominent in annual assessments of the business environment in East Asia.

Economic growth in East Asia has long drawn wide interest in the institutions of South Korea, Japan, China, and Thailand. The effect of international trade and investment on local innovation systems has drawn renewed attention in the scramble to conclude preferential trade agreements in East Asia. Studies of political economy, technology, and offshore investment in Japan offer insight on links between investment and innovation. Patterns of "embedded" mercantilism in the wider economy and of "permeable insulation" in the business community deeply affect the directions of FDI. Studies of techno-regionalism and techno-globalism now complement "techno-nationalism" shaping national innovation systems (Samuels 1994). Debates over Japan's role in regional technology transfer provide markers such as location or "nodes," governance, and institutional structure, for assessing a new role in knowledge networks.

Knowledge nodes

Despite the mobility of investment capital and rapid changes in technology and knowledge flows, more familiar categories of time and place still largely define how we produce, learn, and consume. Chronology and geography provide keys to understanding the formation of nodes. A *node* is the nexus or site of strategic functions within a network (Castells 1996). *Nodes* denote the critical gateways or linkages within global value chains. Finally, nodes in manufacturing chains are grounded in industrial districts or "hubs" across the region with advantages of spillover effects for specialized firms. Japan's chain of production in East Asia has evolved across five decades or so of regional FDI. War reparations in the Treaty with South Korea of 1965 included yen loans for purchase of Japanese technology, and also brought the first wave of Japanese firms to the peninsula since the end of colonization in 1945. The Plaza Accord in 1985 led quickly to an appreciation in the value of the yen, forcing first larger firms, and then smaller firms abroad in the region, dramatically extending Japanese offshore production sites in Southeast Asia, and from the next decade, in China as well. Penetration of global value chains and competition with the West in the region by the 1990s spurred integration of offshore factories into networks, often with headquarters in the region rather than in Japan itself. Only recently do we find the evolution of manufacturing nodes abroad, spurred by the shift from vertical to horizontal networks of investment, and by trade liberalization permitting integration of chains across the region (MERI 2006b: 7).

Why do some urban nodes or "hubs" manage to anchor productive activity while others do not? Japan's own trajectory of technology development provides one answer. Morris-Suzuki wrote not of state funding, but rather of a state coordinating role of "creating nodes in the networks through which knowledge of new techniques could flow to many parts of the industry system (1994: 183)." The state as rulemaker and as supporter of innovation can play a major role in industrial "agglomeration," as do large firms. What we sometimes overlook is the importance of the grounding or "embeddedness" of firms locally and in networks across borders as major factors in the anchoring process (Markusen 1996). Much attention has been devoted to specialized industrial clusters in the study of agglomeration, and now to the linking of clusters across borders in nodes. We can typify hubs or "industrial districts" more generally with proximity, sectoral specialization, and corporate scale. Markusen first enumerated the conventional types of European industrial districts, all variations of the initial Italian model of strong, socially embedded local ties buttressing knowledge transfer among SMEs. She then cited a hub-and-spoke model of large local or transnational firms supporting a network of smaller industries, more familiar in East Asia. One effect of geographical concentration and sectoral specialization would be external economies and knowledge spillovers.

Governance refers to the inter-firm relationships and institutional mechanisms through which non-market, or "explicit" coordination of activities in the chain is achieved (Humphrey and Schmitz 2002)." It includes management by lead firms of key parameters such as product composition, manufacturing process, scheduling and logistics of delivery. Responsibility shifts in time as improvement of local capabilities permits transfer of governance functions to local subsidiaries. Why are some assembly hubs able to move up the technology ladder from simply assembly to parts and component manufacture? Borrus and colleagues argued that global firms in value chains "access locational advantages at each network node associated with the increasingly specialized technology, skills and know-how that are resident there (2000b: 2)."

South Korea remains a model of industrial upgrading in a trajectory of indigenous firms using imported technology to develop into component suppliers, or even end-product competitors in selected sectors. South Korean firms provide components in some Nikkei GVCs, and compete directly with Japanese firms in other sectors such as home electronics. Beyond Korea, market scale and opportunities for local sourcing now draws the majority of Nikkei regional investment to China. Southeast Asia remains attractive for Japanese investors in a "China + 1" strategy to hedge against political turmoil on the mainland (MERI 2006b: 8). Specialization within the region continues to evolve, as Felker noted: "the region's advantages as a node within a networked global production economy … are evolving towards particular niches and horizontal clusters of multinational activity (2004: 101)."

Learning between Nikkei firm and local supplier begins on the shop floor in the distinctive Japanese form of *genba shūgi* or "learning by doing on site." Japanese state, industry associations, and firms cooperate in the governance of technical training. Japanese officials provided core funding for the Thai Textile Institute and the Thai Automotive Institute. Institutes take responsibility for developing national standards, including such ambitious projects as a standard color-coding system for dyeing among textile firms. A variety of state-funded programs post Japanese technicians to local suppliers abroad. Their goal is to train machine operators and management personnel. This model of unit-directional learning can easily be linked to vertical networks based in Japan, with sites abroad limited to assembly of imported parts and components. Localization of parts supply evolves over time with greater knowledge of the local context. Japanese sent technicians abroad first to colonial (1910–45) Korea and post-colonial South Korea, then to Thailand and finally to China. Nikkei firms in time became familiar with networks of reliable suppliers, learned local patterns of labor relations, and cooperated with local states for both technology transfer and foreign investment. A recent shift to horizontal networks of production and supply within manufacturing nodes intensifies the search for talent and sources of knowledge in host nation, whether of market, machinery, or state policy.

What I term "knowledge nodes" within networks of GVCs can serve as strategic locations for innovation through exchange of codified and tacit knowledge. Ernst offered a useful definition of *innovation* as the "skills, knowledge, & management techniques needed to design, produce, improve and commercialize artifacts, i.e., products, services, machinery and processes (2006b: 18)." Tasks would include basic research, applied research, and product development, with the former two types mainly in Japan and advanced industrialized countries, and the latter often pursued as well at manufacturing sites abroad in Asia. Research & Development (R&D) Centers provide one example, but such nodes are also apparent in Standards Institutes or research/training centers associated with specific industrial sectors, as well as at information and planning offices of industry associations.

Conclusion

Production networks and now preferential trade agreements have ushered in a new era of integrated commerce and investment in East Asia. A more sophisticated, more densely articulated division of labor has accelerated the process of market integration across borders. Multinational firms expanding abroad chose sites and established nodes largely on the basis of simple comparative advantage through the 1990s. Organized interests in home and host country have emerged more recently to help shape those decisions. States in tandem with the organized interests of transnational capital in firms and industry associations compete for a role in mapping and managing the evolving division of labor in East Asia. A transition from offshore assembly sites to knowledge networks would appear to benefit home, host, and region. Home country investing firms would benefit from knowledge of market, consumer taste, and local craft abroad. Host country firms would gain from technology, but also marketing and management resources. The region would benefit from a more sophisticated and competitive economic integration.

Improvements in communications and transport contribute today to better integration across national borders and regions. More intense market competition in the New Asia is driving a fundamental change in how business is managed, forcing improvement of production chains, a scramble for knowledge resources, and diverse state strategies of regional integration. Japan's Asian neighbors are making the shift from global production to global innovation networks, pushed by smart governments and organized business interests. Ernst notes that Western competitors are taking advantage with investment and cooperation in product development, but asks, "where is Japan (Ernst 2006a)?" Government units such as JETRO publicly acknowledge that the Japanese economy needs revenue from East Asian investments to leverage its own growth. Revenues from regional investments increased 34.5 percent on an annual basis, to $9.6 billion in 2005. But more important than revenue alone, JETRO officials acknowledge the need to incorporate global knowledge resources. "The time has come to shed insistence on acting alone and to take advantage of technologies, know-how and human resources from around the world (2006a: 31)." Curiously, however, there is little reference to knowledge resources in the region.

Difficulties of communication with local suppliers in offshore manufacturing continue to plague Japanese firms. Some progress is apparent. About two-thirds of firms with offshore manufacturing recognized localization of management and procurement as necessary in a 2006 survey, as against one-third opting rather to "centralize authority in Japan." But even among those pressing for localization, just better than one-half reported satisfactory progress. Problems with the legal system and intellectual property rights protection led the list of localization difficulties in

China (JBIC 2006b). Clearly the preference for intra-firm, and then for ties with other Japanese firms abroad remains strong: "there is still a strong tendency to favor relations with other Japanese firms over firms with different nationalities, even after we allow for knowledge protection, transaction, and other considerations (Ando, *et al.* 2006: 23)." Ernst finds the distinction particularly salient in regional research: "Japanese firms continue to assign a low priority to an expansion of R&D in East Asia (Ernst 2006a)."

Familiar problems for Nikkei suppliers of limited accessibility to technology, slow reaction in structural adjustment programs, and centralized governance come to light in comparisons of US and Japanese firms abroad. Perhaps most disturbing for host country partners, Borrus and colleagues found that Japanese firms gave priority to maximizing value-adding at parent firms in Japan, versus local value-adding (2000b). A recent study of technology transfer and personnel training for offshore subsidiaries of a sample of Japanese and European elevator firms revealed slower transfer among Japanese firms, but also less effective training. Mizuno found that European firms proved more attuned to advantages in the international division of labor by training lower-level personnel more quickly, with Japanese firms investing rather in middle-level personnel (2006).

An interplay of context and coordination in offshore manufacturing help to explain differences in hierarchy and node formation among nations and industrial sectors. Japan's knowledge nodes in leading Asian clusters such as Shanghai, Seoul, or Bangkok tells us of the problems but also of the promise of interactive learning in manufacturing FDI. The differential embedding of Nikkei production in China, South Korea, and Thailand can shed light on how social systems of production coalesce or contend in cross-border manufacturing projects. The stories of Panasonic, Toyota, Itochu, and Toray in the three sites lend a corporate texture to the evolving picture of factory, network, and node. Re-organization of production and learning in three host nations reveals three very different modes of inclusion in the globalization of the New Asia. Regionalism in turn alerts us to changes supporting new knowledge streams in East Asia, a challenge for Japan as well as the West.

2 Insulation versus regional integration

President Kuroda of the Asian Development Bank recently coupled knowledge with prosperity and public policy: "[K]nowledge is the foundation of productivity and competitiveness, and the backbone of good public policy (Kuroda 2007)." This chapter examines the challenge of integrating Japan's national innovation system into regional investment. Will Japan choose to keep innovation at home or to integrate innovative resources across the region? If knowledge is the backbone of regional public policy, integrating the region into Nikkei knowledge networks is the key to Japan's "soft power" in the New Asia.

Storper wrote of *knowledge* as the "institutionalized, embedded social practices, conventions and rules" critical for effective economic coordination (2000: 57). Transfers of productive practices and routines in value chains can benefit both home and host nation. Home country firms moving abroad thrive with supporting industries and new markets. Host country firms gain foreign capital and technology which can contribute to industrial development, employment, and new streams of trade and capital. For firms from the home economy, manufacturing abroad presents a unique opportunity for learning about local skills, labor, and consumer markets, as well as tapping new research resources.

Direct investment in manufacture abroad can support learning in host countries with capital, skills, technology and even product brands necessary for entrée to international production networks. Guyton argued that FDI can also promote local skill creation, and bring new competition to local firms (1996). Technology support institutions are one measure of effective public policy, but productivity ultimately depends on learning at the level of firms.

States and firms

Japan's networks in China, South Korea, and Thailand face a new regional and global competition for adding value. Firms must add significant value at each step in the production process in the effort to link manufacturing with value chains across borders. Nikkei networks have played a major role in East Asia's emergence as the "world's factory" for electronics and textiles, and increasingly for auto vehicles as well. Japanese firms have long prospered with process innovation and product development at home. But today, China's rise poses one challenge to

Japan's leadership in the regional division of labor. Scientific and manufacturing achievements in South Korea pose a further challenge immediately in consumer electronics, and over time in mid-level technologies in general machinery, textiles, and auto parts. Welcoming FDI and tapping foreign know-how for local technological progress, Korea and especially China opted for engagement rather than Japan's precedent of insulation within a national innovation system. At the same time US and European firms have moved to Asia for both production and research, openly searching out new knowledge resources and penetrating local as well as export markets.

Beyond firms, states too have played major roles. Nikkei networks in East Asia emerged with strong state support at home and abroad, whether through direct aid programs, or through the promotion efforts of state offices such as METI and JETRO. Among earlier efforts, Japan's treaty with Thailand in 1955, and a decade later with South Korea brought indemnities in the form of grants and loans for Japanese industrial projects. The Japan–Korea Treaty included $400 million of credits for industrial development, mainly in purchases from Japanese firms (Saneh and Goonatilake 1994). Officials three decades later announced the Miyazawa Plan for Thai SMEs in 1999, a Japanese state project to design and fund new institutions for promotion of "local supporting industries (TN 1999)."

Interventions in South Korea and Thailand pale in comparison with Japanese state efforts to promote economic ties with China. Overseas Development Assistance (ODA) loans to China for infrastructural projects from the 1980s helped to build the energy and transportation structures necessary for profitable private investment. China and the Asian NIEs now play an independent and sometimes competing role with Japan in regional affairs on critical industrial issues such as product standards and regional financial policy. China's proactive political role in the region has undermined Japan's earlier leadership role of "leading from behind" in the economic policies of her major production partners. Japan now promotes regional economic integration in critical fora such as ASEAN + 3, a policy which would strengthen well-established Nikkei networks in the region. Still recovering from decades of political isolation, China without a similar nexus of economic ties looks to a broader political and possibly cultural coalition (Yoshizaki 2007; Zhang 2007).

Knowledge networks

Collaboration in manufacturing across borders has driven regional integration in East Asia, rather than a top-down political strategy among states. Regionalism today provides precedent and suggests prospect for enhancing Japan's regional role. An effort to reshape production ties into knowledge networks would match the initiatives of Western firms, contribute to a deeper economic integration of the region, and lend new credibility to Japan's "soft power." Extending Japan's model of interactive, incremental, continuous innovation into regional production networks would support two-way knowledge flows between home and host firm, whether at subsidiaries, affiliates, or sub-contracting independent local firms.

Research abroad to adapt products, but also market and consumer research for developing and focusing brands provides other paths of knowledge exchange.

Elusive goals of regional production and product standards represent yet a further and very fundamental knowledge flow. This would include discursive standards such as International Organization for Standardization (ISO) certifications for management, or sectoral standards for product quality. Benefits might include more reliable local production, lower costs for procurement, and improved access to local labor and consumer markets for Nikkei firms. Host country firms would gain not only in knowledge of craft, international standards and markets, but also augment their resources for anchoring increasingly mobile international investment capital. The goal is not convergence of technological levels, but rather enhancing comparative advantage for home and host firms.

Learning strategies differ according to a nation's technological capability, industrial sector, scale of enterprise, and orientation to local or export market. Some Nikkei firms have already moved down the path of localization, but regular surveys of Japanese enterprise in the region indicate less extensive local procurement and outsourcing than among their US competitors (JETRO 2007a: 31). What might explain such reluctance?

Fears about patent infringement, leakage of confidential information, or weak enforcement of contracts, particularly in China, reinforce insulation of innovation networks. Low levels of educational achievement in Thailand, an inadequate supply of engineers, and unreliability of local production deter some Japanese producers from moving beyond the tight circle of Nikkei suppliers. Finally, intense competition with South Korean firms, as well as their proven record of ingenuity in technological development would not encourage mutual learning through direct investment without reliable protections. Such factors encourage further insulation rather than more cooperation in innovation strategies.

Coordination and context

The term "public good" implies that it is neither feasible nor desirable to exclude a third party from using a concept or design, even though they may have made no investment in creating it (Maeda 1999). Knowledge is fundamentally a public good that circulates through education, print and electronic media, and the business world. Constant analysis and then reanalysis of social practices in market, polity, and society distinguish modern society linked by new modes of interaction. A continuing revolution in communication and information technologies promotes a "double hermeneutic" in which concepts are continually reshaped through reflection and interaction (Giddens 1990). Moreover, the critical resource in technology transfer is usually "tacit" knowledge such as know-how, rather than information that can easily be codified in manuals. Know-how, in turn, can only be communicated through the experience of interaction with knowledgeable people and institutions.

Free-flowing streams of knowledge appear critical to insure creativity from diverse sources. A broad-based exchange of technical information, for instance, appears a necessary condition for efficient development of new technologies. But what we find instead is a curious process of bordering in knowledge flows across firms where the acquisition and proprietary use of knowledge through patents, licensing and royalties has helped to drive growth in manufacture as well as information industries. The same intellectual property rights that provide incentives to create new knowledge can also impede socially optimal exchanges. Ironically, the amount of resources devoted to "boundary-setting" pales in the face of the huge investment in "boundary-maintenance." Difficulties in enforcement of boundaries on knowledge through patents and licenses have spawned an industry of legal experts. Shoring up knowledge "spillovers" or unintended transfers is just one example (Kohpaiboon 2006). Innovation is thus a curious combination of cooperation and competition.

Innovation begins with technological capability, often acquired through hands-on trial and error, but accelerated through access to advanced technology, know-how, and widely recognized standards. The immediate goal is acquisition of specialized knowledge. Cooperation with other firms and institutions permits access to complementary stocks of knowledge which, in combination with specialized expertise, can lead to process improvement or product development. Two features of innovation in the Asian NIEs proved critical: the systematic or organized character of the effort, and the dynamism or speed of the process (Altenburg 2006).

Other features of especially Japanese innovation include scale and scope. *Scale* denotes the size of enterprise, with larger and smaller firms both playing critical roles within the *keiretsu* vertical organization of enterprise. Specialized SMEs within industrial districts would be assigned specific responsibilities for developing and then producing new products. *Scope* refers to the range of products produced. A larger firm in Japan can extend its product line by tapping the specialized expertise of smaller sub-contractors specialized in specific products.

Japan's model of cooperation, whether among larger and smaller firms or within industry associations, may prove its most enduring contribution to the development trajectories of East Asian nations. The dominance of SMEs in Thailand's supporting industries, the research resources of South Korean SMEs, and the demand for more reliable production among China's SMEs have driven state and industry efforts in the region to improve knowledge flows for smaller enterprise. But productive local linkages between larger and smaller firms familiar remain problematic in the three host nations for Nikkei investment.

Scale also affects solidarity in industry associations. Associations often prosper in concentrated industries with a relatively small number of major firms. Industry in both Japan and South Korea has evolved with a few large firms dominating sectors. This contributes to parallels in the form of Korean and Japanese industrial associations, but often not in function. One also finds concentration among state-owned enterprises in China's major industrial sectors, but with industry associations only recently spun off from government offices unaccustomed to coordinating and representing a "sectoral" interest distinct from government policy. A smaller scale of enterprise in Thailand's "supporting industries" does not prove fertile ground for sectoral associations. The few industrial sectors with larger local firms in Thailand, such as spinning or weaving, have provided a base for stronger associations, though these remain the exception rather than the rule. Yet scale in all four nations is but a single factor in the formation and function of associations. Closer analysis of social systems of production provides a far more comprehensive understanding of how and why associations develop.

Trust and social capital

The term "scope" also suggests the breadth of knowledge exchange in the Japanese ideal of *monozukuri*. It connotes craft and specialized know-how, and denotes the making of things with skilled, hands-on human labor. Craft on the factory floor proceeds with extensive knowledge flows in a constant interchange of information and skills. This coincides with recent attention to learning within a "collaborative community." Executives from home country firms must learn to tap all available knowledge on the host country factory floor from staff experts to machine operators to insure continuous improvement. What provides the basis for effective exchange of tacit knowledge is trust between home and host partners, essentially a social rather than solely technological interaction (Adler and Heckscher 2006). Shifting borders between the technical and the social have led some to term the institutions of innovation a "socio-technical system (Kline and Rosenberg 1986; López-Martínez and Piccaluga 2001: 278)." Needs and tastes of end-users provide the goal for technical design and production, while legal, market, and financial systems provide boundaries. Such interactions, intensified in the reflexivity of contemporary society, highlight the communal context of learning.

Production networks across national borders demand new types of solidarity to insure trust and facilitate communication. Traditional solidarities based on locale, ethnicity, or nationality become problematic in the construction of global networks linking diverse communities. Norms of trust within networks represent a type of social capital supporting collective action for mutual benefit (Woolcock 1998). Two aspects of social capital appear central to the transition to knowledge networks: the "intellectual" character, and necessary boundaries. Social capital develops with shared representations and systems of meaning critical for knowledge networks. Personal ties, often supported in Japan by institutional links as well, permit an intellectual capital or "knowing capability" for an organization or professional practice (Nahapiet and Ghoshal 1998).

But the boundaries of such societal resources may impede their utility due to shifts in state policy or social patterns which erode initial agreements. James Coleman's earlier conclusions about the "trustworthiness of the social environment," and the stability of social structure appear even more problematic in global or even regional society across national borders (1990). Changing investment policies, inconsistent enforcement of laws, and a lack of transparency continue to hamper investment across borders. More importantly, the instability of states and societies in East Asia, due in part to uneven trajectories of political democratization and market liberalization, erodes the trust or deeper social capital developed within Nikkei networks of investment and production. Firms moving abroad must strike a balance between exclusivity or insulation, and flexibility or openness to new members and new ideas.

Electronics firms can move abroad with modular assembly based on standard interfaces. Communication of assembly procedures remains relatively basic with initial assembly of very simple products, but intensifies with more sophisticated modules and end-products. Textile firms often establish midstream spinning firms at the outset, and then move upstream to synthetics once a base of communication and engineering expertise abroad has been established. Trading firms generally start further downstream in the value chain of textile production with labor-intensive assembly of garments, but often reinvest over time in fashion with communication of designs and more sophisticated cutting and sewing. Perhaps no major sector demands more intensive exchange of information on design, fabrication and performance in Japanese manufacturing than auto assembly.

Trust in supplier relations develops across years of product development, manufacture, and process improvement. Trust evolves as the project demands more interdependence, clearer norms and better reliability, and mutual interest in deepening cooperation (Holzhausen 2000; Smitka 1998: 165). Context plays a role as well as contract. Societal context helps to shape not just norms and reputations, and may provide structural incentives for cooperation to discourage cut-throat competition.

What happens then as we stretch that context across borders? A global search for ever lower supplier costs, coupled with standardization of parts and components has forced Japanese carmakers to source beyond their traditional networks and well-established local bonds of trust. A new, more open model of collaboration between core engineering and design firms and their suppliers affords firms more explicit definition of norms and expectations, and specific guidelines on cooperation, but offers no substitute for context-specific bonds of trust (MacDuffie and Helper 2006).

Innovation in Japan

Innovation at leading Japanese firms has been widely applauded and with good reason. Patent activity offers one measure of innovative activity. Panasonic ranked second in the world in 2005 with over 2,000 applications for patent recognition to the World Intellectual Property Organization (WIPO). Toyota Motor ranked nineteenth with about 400 applications, and Japan's leading technology support institution, the Japan Science and Technology Agency, ranked thirty-third with 300 applications (WIPO 2006b). Larger firms with huge research budgets can flourish in the patent competition, but what of the smaller firms?

Contributions of Japan's smaller firms often get lost in discussions of industrial dualism or "gradation (JSBRI 2006)." Certainly the fate of smaller Japanese firms is closely linked to the success of the larger firms, usually as a long-term sub-contracting partner in a *keiretsu* or group of enterprises. A dual industrial structure of larger and smaller enterprise, some suggest, results in a dependence on larger "parent" firms. Some conclude that smaller firms survive only with loyalty and reliable assembly, with little technological contribution. One author even suggested that SMEs can afford to be technologically "less aggressive," and simply cooperate with innovative (larger) partners in a division of labor permitting them to concentrate on production (Okada 1999). But others question the premise of SME technological dependence on larger enterprise. Morris-Suzuki argued that the misplaced emphasis on state-led national innovation projects diverted attention from the core dynamic of social networks of innovation, particularly strong among local SMEs from the late nineteenth century. This "bottom-up" innovation facilitated adaptation to technological change, and development of know-how in the "supporting industries" of Japan's modern industrial sector (1992; 1994).

Even the state did not recognize the innovation among SMEs in the recovery decades following World War II. Only recently has the government looked to the role of unaffiliated or "independent" SMEs in national innovation and employment (Ueno, *et al.* 2006). A significant result of this bottom-up process of innovation was its role in helping small firms to adapt to rapid technological change, and in many cases to become sources of parts and supplies for the modern industrial sector, setting a precedent of "supporting industries. (Morris-Suzuki 1992)."

National innovation system

There is little doubt that Japan is making every effort to retain her technological superiority in East Asia. State policy and programs, implemented in tandem with technology support institutions, have provided overall direction for technological development since the 1950s. An emphasis on self-reliance from the 1960s superseded initial imitation of foreign technology in the years of recovery from World War II. Japan refocused technology policy on indigenous innovation and basic research by the mid-1970s. The results are impressive. Japanese researchers led the world in patent applications in 2004, and ranked second only to South Korea in the ratio of total applications to amount of R&D expenditure. Among sectors, transportation and automobiles ranked among the most heavily funded areas, as well as electronics and electrical machinery. Machinery for telecommunications, automobiles and electronics together accounted for 48 percent of total R&D that year (MEXT 2006b). Government and private sector continue to lead the world in per capita investment in R&D, amounting to 3.1 percent of total gross domestic product (GDP) in 2005, and trail only the USA in total research expenditure and number of patent applications.

Smaller firms face the dual challenge today of innovation and independence. Survival and success in global competition demands innovation. Diversifying contracting beyond the single larger partner permits greater flexibility and wider access to knowledge flows. An official ideal or "vision" of smaller firms with core manufacturing technologies, and larger firms focusing on final products still dominates in Japan. The major change from the postwar vision is the role of SMEs in innovation, usually in tandem with larger firms for development of "cutting-edge strategic products (JSBRI 2006: 89)." What remains central to the innovation system are the social networks that develop among the producers, particularly strong in Japanese industry (Granovetter 1994).

Unlike more informal kinship networks familiar across Asia, Japan's networks are rooted in overlapping institutional ties, as evident at cooperatives and industry associations within local clusters (Imai 1992). The term "supporting industries (*susono sangyō*)" denotes a network of SMEs in support of larger firms, often in a *keiretsu* form of vertical integration. METI recently projected the term across the industrial structure of the region "in which support for industry overlaps between countries, creating a mountain-like structure of supporting industries (METI 2007b)." A growing demand for local input in product design and development abroad provides an opportunity for new knowledge flows with local firms. Strategic alliances with leading local firms, such as Sony with Samsung, or Sanyo with Haier, have already pioneered new knowledge flows in specific products. Expanded local market research and product development within affiliates abroad likewise represents new knowledge flows. Upgrading local SME supplier networks abroad poses a far different challenge, but also a far broader range of possibility.

Cooperative learning between larger firms and smaller, specialized firms has played a major role in Japanese innovation, often in exclusive inter-firm networks which insure trust (Okada 2001). More independent SMEs, as well as efforts of large-scale enterprise to diversify supply networks to reduce costs among major assemblers, have permitted wider participation at home in the new era of "meshing" with multiple clients and suppliers, which reduces dependence on major customers. Does diversification of partners reduce levels of trust for exchange of critical information? Government surveys highlight two conditions for sustaining information flows: a mutual commitment to consistent product quality, and active involvement in technical exchanges (JSBRI 2007). But can large Japanese firms abroad both mesh and maintain information flows with local SME networks abroad, particularly in areas of integral manufacturing where Japanese firms excel?

Cooperative learning

Cooperative learning implies both process and result. Process includes structures for cooperation within firms, industries, and associations that can insure trust. One comparative study of process in US and Japanese firms found the latter strong in the codifying of tacit knowledge, but weak on sharing tacit knowledge across boundaries within the firm, or in combining codified knowledge from multiple sources within and beyond the firm (Motohashi 2003). A recent study of organizational reforms at Panasonic from 2004 highlighted how strategies were formulated to address both weaknesses (Kodama and Suzuki 2007). The result is the ability to learn or absorb "tacit knowledge" across regional or national cultures, whether among individuals or the wider society.

We can look to process and result in various areas: a) in operations on the plant floor, b) in market and product research at firm or industry institute, and finally c) in cooperation in industry standards and perhaps regional sectoral policy. Machine operators must learn how to manage installed machinery, adapt machinery for new products, and calibrate machines to improve productivity. Engineers and researchers need to develop new or improved manufacturing processes, and new products, often through research on product development for the local market (Rodrigo 2001). Efforts to establish and enforce common standards of management, production, and marketing represent a third, critical level of cooperative learning.

Each level of the process taps different flows of knowledge. Learning within local SME networks begins with production methods and may stretch to market or product research. Knowledge networks within the hierarchical structure of mother firms and affiliates evolve with operations and product improvement, and possibly market research. Strategic alliances with major local firms in the sector would be limited to research in a specific product area, and cooperation on industry standards.

Affiliation in vertical *keiretsu* remains a critical institutional tie in Japanese innovation. Long-term ties with the same clients provide a rich interchange of information on market and technology, but also opportunities for collaborative R&D (Okada 1999). Among core *technology support institutions* today, universities and technical colleges, local/regional technical institutes, and industrial associations each play different roles for firms, depending on scale, sector, and affiliation. SMEs in regional clusters benefit from information exchange and training at local and regional technical institutes, cooperatives, and business associations. Universities and colleges provide more advanced R&D support for smaller, "independent" firms (i.e. without *keiretsu* affiliation), and specific project support through contract research for larger firms. Yet in most cases, a more informal pattern of personal ties of association or university researchers and firms contrasts with the formal, contractual patterns of university and association links to industry in the USA (Kodama and Suzuki 2007).

Cooperatives and sectoral associations function more as technology facilitators and as sources of information for firms large and small, rather than for actual research. Yet their primary role of building consensus on industry directions, and lobbying government on sectoral policy plays an indirect role in promoting technology and gaining research support. A pyramid of business representation in Japan includes local area and sub-sectoral associations, sectoral or industry associations such as Japan Automobile Manufacturers' Association or the Japan Chemical Fibers Association, and then peak-level groups such as the Keidanren and the Japan Chamber of Commerce and Industry (Lynn and McKeown 1988; Tachiki 2005).

Architecture

Beyond process and results in learning, patterns of assembly or "fabrication" deeply affect the content and intensity of knowledge flows in innovation. Standardization of parts, components, and even modules encouraged the segmentation or specialization of supply chains from the 1970s. Standardization in turn accelerated the trend among MNEs to tap new resources of labor and knowledge worldwide, moving production to the most cost-effective sites. More recently, the "off-shoring" of services as well as production by US and European firms has brought new MNEs to

Asia in search of innovation resources. But standardization differs greatly according to the type of product.

A module can be designed with a single function and a standard interface to insure compatibility with other modules. The interface makes possible interdependence without compromising the single "independent" function of each module. This permits a manufacture of modules acquired from a variety of sources, and then brought together according to the proprietary design of a lead firm. Yet even within sectors, some segments can be modularized, while others remain more specialized, demanding customized know-how.

An integrated production process in textiles provides one example, with fibers moving to fabric through spinning and weaving, and then to apparel with the cutting and sewing of cloth. Chemical fiber production upstream and high fashion downstream demand more integral, specialized know-how, but midstream spinning and weaving, and commodity apparel downstream permit efficient modular production. Modular architecture of continuous production in Japanese spinning and apparel facilitated its transfer to Southeast Asia already from the 1960s. State and cooperating *jaebeol* or family-owned conglomerates in South Korea prospered with an indigenous line of modular production in apparel. They then licensed Japanese technology in chemical fibers from the late 1960s to move upstream into chemical or "synthetic" filaments and fibers.

Production of consumer or "household" electronics permits even more compact and mobile modules. Modular production of consumer electronics such as TVs, refrigerators, and personal computers initially caught the attention of scholars with its rapid development in East Asia from the 1980s. The transfer of know-how included mainly calibration and operation of imported machinery, with specifications and designs imported as well. South Korean and Taiwanese assemblers contracted with leading foreign firms for "original equipment manufacture (OEM)," producing goods then re-branded and sold by those firms abroad. Foreign firms held the knowledge of design and marketing, and the organizational expertise of integrating design and engineering, as well as supply and demand across borders. Asian assemblers took advantage of this knowledge flow to develop their own basic technologies in the industry, and then reinvested in research to move upstream with their own indigenous firms. Chinese electronics firms such as Haier, established only in 1984, followed a similar trajectory in a much shorter time frame.

One key to the global expansion of modular production has been the development of common measurements or "specifications," and standards. Product standards within a sector become critical for the emergence of national and regional industries. Production and sales of dyed fabric, for instance, depend on a common definition or code of colors in the dyeing sector of the textile industry. Establishing an industry standard can prove an advantage for leading firms. Similarly, market leaders in electronics will specialize in design to establish a privately owned but openly shared "architectural standard" for the industry. They will then subcontract out other steps in the value chain on the basis of their privately held standard. Some have argued that this supported a shift to specialization within modular production at US firms. The adjustment helped US firms to regain the lead from Japan in some areas of electronics production from the 1990s (Borrus 1999; Motohashi 2003; Sturgeon 2002). Whether an accurate assessment or not, the competition helps to explain the intensive effort of Japanese state and industry offices to establish product and production standards in East Asia.

If modular production offers new opportunities for market leadership among home country firms moving abroad, it also opens opportunities and risks for host countries and their innovation systems. South Korea has excelled at unbundling the technology packages in OEM, often with the help of contractors and their inspectors from abroad intent on upgrading the quality and reliability of offshore production. But labor-intensive assembly of imported, standardized modules elsewhere has sometimes fostered the "deskilling" of labor. Workers with a variety of local skills are refocused on a very basic set of assembly tasks with endless repetition. Thailand has been less successful than Korea in upgrading local innovation, as evident in the limited number of competitive local firms in electronics and textiles.

Modular production in electronics, textiles, and other sectors has led one scholar to term China "the world's factory of modular production (Ohashi 2006)." The rapidity and scale of inward FDI in Chinese manufacturing sectors does not yet permit a clear profile of benefit for local innovation. A vigorous debate has ensued over the skilling or deskilling of labor in the largely foreign-invested assembly industries of China (Lemoine and Ūnal-Kesenci 2004; Sturgeon and Lester 2003; Wang 2006).

Integral product architecture

Assembly of standardized interfaces and single-function modules can often suffice for production of apparel or less sophisticated home electronics. Assembly with less common, more specialized types of interfaces makes possible integration of more diverse modules. This "integral product architecture" demands continuous human communication throughout a process of joining specialized parts and components, with a defined but not generalized or widely standardized interface. Automobile manufacture, for instance, includes 15,000 or more individual parts. Many of these can be joined into modules for specific models, but often the architecture of individual modules does not permit integration into a common platform. Major auto assemblers such as Toyota have pressed ahead with limited modularization and even development of common "platforms," but the complexity of the models precludes the level of standardization noted above in textiles/apparel or especially in electronics.

Japanese firms have prospered with modular production in consumer electronics, midstream textiles and apparel. At the same time, enduring features of industrial organization in the home islands such as close-knit networks, long-term employment, and the seniority system, as well as the vertical integration within *keiretsu* have long supported integral manufacturing (Motohashi 2003; Ohashi 2006). Nikkei firms continue to dominate more specialized, integral manufacturing segments such as proprietary design or critical components, even within wider lines of modular production. Ohashi cited "light, slim, very compact (*keihakutansho*)" consumer products in electronics as an example of Japanese integral manufacturing. Japan supplies the more expensive, critical parts for electronics exports from neighboring South Korea and China. Japanese components often represent the higher value-added parts, even in a relatively sophisticated industry such as Korea's consumer electronics exports.

Integral manufacturing has proved feasible within Japan's network society, and indeed profitable for Japanese firms, although critics question the viability of a "vertically integrated business model of self-sufficiency" given that standardization is proceeding rapidly even in traditional integral industries such as auto production. Yet a recent JETRO White Paper on International Trade would suggest that an integral business model remains the nation's competitive advantage. Officials recognize that firms have depended primarily on "integral production systems" with most design and manufacturing in-house, and cite great caution among firms about modular-type manufacturing that requires sourcing beyond the firm.

One solution would be modular production or "outsourcing" for general use products under a Japanese brand name. But the more intriguing suggestion is for integral type "technology" for general-use products. In other words, maintaining the networks and the sharing of production within the firm and close circle of suppliers, but increasing reliability and lowering costs by closer coordination. One example would be the sharing of concept and design development tasks with auto parts manufacturers from the start of a new car project.

Integral production systems demand close cooperation. Scholars of network theory remain divided on the advantages of solidarity versus the disadvantages of insulation. A long tradition of cooperation within and among Nikkei firms may impede the flexibility necessary to move beyond the borders of the Nikkei networks to capture knowledge resources of host country firms, large and small. Codified knowledge in more open US modular networks, unimpeded with deeper contextual roots or extensive exchange of tacit knowledge, may permit a more competitive flexibility in areas of rapid technological change (Sturgeon 2002).

Target firms

Five global corporations in Japan offer insight into practice and potential for the shift from assembly to knowledge networks (Table 2.1). Industry leaders in sales, production, and research, they are also sectoral leaders as investors in E[illegible] Prominence in industry associations permits a voice as well in state pol[illegible] abroad, Japan's "soft power." Each of the firms is deeply involved in product, process, and/or market research. The five represent the diversity of enterprise organization in Japan today. Toyota Motor Company and Panasonic are major, independent horizontal groups, although Sumitomo *keiretsu* firms hold a 13 percent stake in Panasonic. The Toyota Group includes three automakers, Toyota Motor, Daihatsu and Hino Motors, as well as parts suppliers such as Denso, and a trading firm, Toyota Tsusho.

Suppliers take part in planning, design, and initial engineering of products in

Table 2.1 Financial performance for selected firms in 2007 ($US billions)

Firm/Data	*A: Sales*	*B: Percentage overseas*	*C: Net income*	*D: R&D expenditure*	*E: Employment (% overseas)*
Panasonic	96.8	48%	2.81	5.54	305,828 (56%)
Toyota Motor Corporation	262.8	52%	17.1	9.5	316,121 (78%)
Denso	40.1	50.8%	2.4	3.1	118,853 (69%)
Toray Industries	16.4	41.6%	0.48	0.45	38,565 (55%)
Itochu	124	42%	2.18	n.a.	48,675 (92%)

Sources: TKS (2008); Panasonic (2008c); TMC (2008); Denso (2008d); Toray (2008a); Itochu (2008d).
n.a. = not available

Period: April 1, 2007 – March 31, 2008. US dollar amounts computed at the rate of 100 Yen = US $1, the approximate current exchange rate on March 31, 2008.

Japanese innovation strategies. Panasonic, a leader in home electronics with brands such as Panasonic and National, pioneered a more open procurement network of independent suppliers in electronics, as opposed to the more conventional, vertical organization of major *keiretsu* (Table 2.1) (Lincoln 2006). Auto assemblers such as Toyota gather first-tier suppliers into *kyōryokukai*, or *keiretsu* "cooperatives," to join in planning and design of new products. Toyota jostles with General Motors for the lead among global automakers, and with Panasonic in Japan for the most annual patent applications.

Toyota Motor Corporation and the affiliated Denso have extensive production facilities in China and Thailand. Suppliers learn of management systems and strategies, gain access to technological services, but remain themselves responsible for research and design. The requirements of capital and engineering resources, as well as quality control preclude participation at this level of all but a few major first-tier suppliers (Ohmori 1999). Denso is Japan's leading auto parts firm, specializing in electric and electronic components, and boasts the largest research budget. Toyota Motor is the leading shareholder, with a 20 percent stake. First-tier suppliers such as Denso will organize a similar association of second-tier suppliers, less for "design-in" tasks than for information sharing, and limited technological services.

Textiles and fashions reveal still other paths in cooperative learning. Japan's top textile and fiber producer, Toray, manufactures in all three of Japan's production partners, with varying levels of technological cooperation. Toray is a member of the Mitsui *keiretsu*, one of the oldest and largest of Japan's six vertically integrated industrial groups (Brown & Co. 2001). Turning from a maker to a merchant, Itochu is the textile and apparel leader among Japan's fabled General Trading Companies, and a pioneer in Nikkei investment in Asia since the 1960s. Itochu is a member of the DKB Group, the most recent of the six large *keiretsu* in Japan. A trading firm such as Itochu coordinates a production stream between suppliers and clients, moving product, specifications and know-how from upstream synthetics producers down to end-users in department stores and boutiques. Smaller firms play a particularly prominent role in textiles.

I review the industry leaders with close attention to home and host country SMEs, the backbone of regional production networks (McNamara 2006). Manufacturing SMEs in Japan include firms with fewer than 300 employees, and a capital of up to ¥300 million (or about $2.5 million). Excess capacity, employment, and debt have thinned the ranks from 5.3 million SMEs in 1986 to only 4.3 million in 2004, and exits continue to far outnumber new entries. Yet smaller firms retain a significant role in the economy. Some 273,000 SMEs in manufacturing employed 5.8 million workers or 71 percent of total manufacturing employment in Japan, and produced half of the total value of shipments in manufacturing in 2004 (JSBRI 2007). They captured 89 percent of the value of domestic shipments in textiles and apparel (¥4.4 trillion), and 15 percent (¥7.7 trillion) of shipments in transport machinery in 2003. Apart from employment and production, the smaller firms are playing an increasingly important role abroad. Less than 200 or about 4 percent of Japanese SMEs registered offshore subsidiaries in 1986. In 2002, the total grew to 433 SMEs operating offshore subsidiaries. SMEs have moved mainly to nearby countries within East Asia, with a total of 1,141 subsidiaries registered in Asia by 2004 (Table 2.2).

Insulation of innovation

Japan's leading official on technology policy recently singled out "continuous innovation" as critical in the nation's goal of becoming the "world's top manufacturing nation (Kosaka 2007)." Highlighting the role of agglomeration or "clusters" in learning, he lauded the "Knowledge Cluster Initiative" for promoting local collaboration within regional clusters. Geography and sector, a particular location and a specific industry or craft demark the process of "clustering." Traditions of concentration and community represented in associations help to reduce uncertainty and transaction costs, particularly among SMEs. Japan's *sanchi* or regional

Table 2.2 Subsidiaries of Japanese manufacturing SMEs in East Asia, 1994 and 2004

Location/Year	*1994*	*2004*
ASEAN 4 (Thailand, Malaysia, Indonesia, Philippines)	139	382 (33%)
4 NIEs (ROK, ROC, Singapore, Hong Kong)	166	276 (24%)
China	140	483 (42%)
Total East Asia	*445*	*1,141 (100%)*

Source: JSBRI (2006: 82).

* The four NEIS (i.e., Newly Industrializing Economies) include the Republic of Korea (ROK), or "South Korea," the Republic of China (ROC) or "Taiwan", as well as Singapore and Hong Kong.

concentrations of producers in the same sector were particularly prominent in weaving and dyeing of textiles, but continue today as a base for machinery and machine tools, and are important for auto and electronics sectors.

This picture of a tightly integrated, insulated Japan content with process innovation strikes others as dated and inadequate. For instance, Prime Minister Abe assembled his own "Innovation 25" group, beyond the normal structure of ministries, to envision new directions for innovation through the year 2025. Unlike METI Minister Kosaka, Innovation 25 strategy council leader Kurokawa faulted the Japanese tendency to "go it alone" with only craftsmanship and continuous improvement (*kaizen*) as being inadequate for the twenty-first century. Collaboration meant seeking partners who could help to pioneer not simply incremental change, but new modes of thinking (Kurokawa 2007). Critics will not be satisfied with simply incremental change in the nation's technology future.

Insulation

Attention to the "insulation of innovation" parallels an earlier thesis of "permeable insulation" depicting Japan's distinctive response to the challenge of global market liberalization. Officials appeared to insulate the economy of uncompetitive local industries from foreign competition, but liberalize markets and outward investment for globally competitive industries (Schaede and Grimes 2003). Some would suggest that a sunset industry such as textiles survives in Japan only with tariff protection in some segments, and "behind the border" protection of business practice for uncompetitive midstream and downstream firms (Kim Dohyung 2007).

Yet liberalization of investment laws and Japanese government pressure to open markets in ASEAN and China have supported the extensive global networks of Toray. Import liberalization for apparel has also benefited Itochu, now responsible for a large part of Japan's fabric and clothing imports from China. Permeable insulation helps to explain the uncharacteristically *ad hoc* and pragmatic character of Japan's trade policy, particularly for manufacturing industries in transition. But more importantly, it highlights the conflict of interests between a state intent on maintaining a national industrial base, and globalizing firms, often represented by strong industry associations, focused on shareholder profit (Schaede 2000).

Insulation of innovation must not be confused with the original thesis of permeable insulation in Japan's wider political economy. Insulation of the national innovation system keeps basic research and leading product and process research within the Japanese islands. The relative absence of foreign manufacturing firms in Japan contributes to this "go it alone," solitary approach, where a firm relies mainly on in-house tacit knowledge to build up competitive advantage (Motohashi 2003). Japan's outward FDI stocks of 8.5 percent of GDP dwarfed inward FDI stocks of only 2.2 percent of GDP. Even apart from inward and outward FDI leaders like the USA and Great Britain, neighboring South Korea's greater receptivity to FDI and foreign firms has permitted joint-research and patent activities on the peninsula. Non-residents filed only 9 percent of patent applications from Japan in 2005, compared to 27 percent in South Korea. METI officials took note recently also of the low ratio of foreign patent applications from Japan in comparison with Western competitors (METI 2006b: 23; WIPO 2006a). The insulation is not permeable in the sense of penetration by globalizing Japanese or foreign firms. On the contrary, selective engagement through licensing and alliances with technology leaders abroad only reinforces the protective umbrella. Japan reported $13.6 billion of payments for royalty and license fees in 2004, as well as $15.7 billion in receipts (World Bank 2006a).

A recent government review of Japan's technology policy highlighted the changing profile of the nation's trade in technology. As the trade values indicate, "insulation" does not imply "isolation" from international sources and markets for manufacturing technology. Manufacturing continues to dominate the trade, including 98 percent of exports and 87 percent of imports (MEXT 2006a). One major change is growth in annual export values 2002–2006 (Table 3.2), together with incremental growth in imports. The USA remains the leading destination, absorbing twenty-two percent of total value of Japan's exports, with Asian destinations drawing a growing share (China 14%, South Korean 7.8%, and Thailand, 3.5%).

Much of the export trade is directed to Japanese affiliates abroad, maintaining a technological edge within production networks. Of greater interest for insulation is import of technology from abroad, which has not grown in tandem with exports. Nearly three-quarters of all technology imports in 2004 came from the USA, and the rest from Europe. The import profile highlights comparability in technological levels between the USA and Japan, exports from technologically sophisticated Japanese affiliates in the USA, and the close linking of the two economies in trade and protection of intellectual property rights (Smitka 1991).

Innovation policy

Unlike the *ad hoc* approach of permeable insulation, the insulation of innovation has been crafted, coordinated and closely monitored by the state under its national innovation policy. Precedents of foreign technology acquisition and improvement from the late nineteenth century continue to shape technology policy in Japan today. Japan's processing industries originated in a context of national development with foreign firms at quite a distance (Morris-Suzuki 1992). While this "national" experience may render Japan's experience less relevant for internationally engaged developing nations today, it may also complicate Japan's own globalization process. Despite the frequent lament in official policy papers from METI and JETRO about the need for ties with leading firms from abroad in technology development, action plans remain limited to tapping foreign expertise for high-priority "national" technology development (Corning 2004).

A further limitation is the narrow focus on technology leaders, rather than an open search across the world for relevant new knowledge in production and product. Quite apart from sensitive areas of high technology, the insulation even hampers product development in East Asia as Japanese firms compete in new consumer markets. A call for "acquiring global technologies, know-how and human resources, rather than focusing excessively on home markets", in a recent JETRO report was coupled with criticism of "the not invented locally stigma (JETRO 2006a: 36)." A METI Strategy Report highlighted the contradiction between an inward-looking innovation strategy and the broader challenges of adjusting abroad: "It is necessary to overcome problems such as over-dependence on the domestic market, pursuit of Japanese-style operations abroad, the lack of global personnel management, and the lack of strategies for technology, intellectual property and standards (METI 2006b)."

A final feature of the insulation of innovation parallels emerging tensions between state and globalizing firms in the original thesis of permeable insulation. Global market competition in high-risk and high-cost areas of consumer technologies has forced alliances with firms beyond Japan's national boundaries. Electronics giants Sony of Japan and Samsung of South Korea agreed already in 2004 on joint-product of seventh-generation LCD displays for large TVs (CNET 2004). Denso and Toray use Korean subsidiaries with leading technology to establish subsidiaries in China in a new triangulation of investment and technology adaptation. Japanese firms long accustomed to cooperation locally across sectors in technological development, appear now ready to ally with firms abroad to keep pace with the competition.

The speed and cost of technology advances may erode the advantages of Japan's innovation insulation before the state finally moves to promote interaction across borders. Perhaps more urgent is the task of tapping human resources and emerging consumer markets in the region to balance the penetration of global value chains from Western countries. A third factor in favor of interaction is the opportunistic behavior of Chinese and South Korean firms in the global search for a technological edge. Strong national innovation systems in both nations may well force greater Japanese cooperation to maintain its initiative and credibility in regional production networks.

Conclusion: governance in global value chains

Inter-firm ties and institutions contribute to "explicit" or non-market coordination of economic activities. The process of *governance* in GVCs begins when "one firm coordinates the product, process and scheduling/logistics parameters followed by another firm (Humphrey and Schmitz 2004)." Control by headquarters or the lead network firms provides the focus, and sheds light on significant differences between relatively insulated Japanese networks versus more open Western networks (Borrus, *et al.* 2000b; Borrus, *et al.* 2000a). But the study of GVC governance in East Asia demands comparable attention to the role of the state beyond the network. Whether in state economic policy, state innovation policy, or in regionalism itself, the state plays an organic role in the governance in Nikkei networks, in home and host country.

State role

Some scholars laud the "catch-up product cycle" under Japanese leadership. Others decry the insulation and exclusivity of Japanese supply lines. But no one can ignore the close coupling of state and capital as Japanese firms move abroad. Some suggest "mercantile realism" to describe economic competition among states to nurture "national" firms at home but also abroad (Samuels and Heginbotham 1998; Yoshimatsu 2003). China's rise, competition from regional NIEs, and particularly the ability of Western competitors to integrate value-added resources in the region into their chains poses a challenge to Japan's strategy of soft power or "leadership from behind" in Asia.

The Japanese state has quietly pursued a coordination of a division of labor and a "division of knowledge" in East Asia, in tandem with business interests at firm and association levels (Doner 1997; Morris-Suzuki 1992). An effort to harmonize trade procedures including sectoral standards, to facilitate foreign investment, and now to establish a common framework for regional financial stability continues to distinguish state support for regional investment (Hayashi 2006; Yoshimatsu 2006). Expanded training and technology support from institutions such as the Asian Productivity Association, and the shift to the management and technical infrastructure in priorities of Overseas Development Aid (ODA) brought that role to a focus in the past few years.

Officials of the Japan International Cooperation Agency (JICA) declared a new "catalytic" ODA role in 2003. Projects "designed to bring economic benefits to Japan" would include: 1) "those aimed at promoting free trade agreements, 2) strengthening economic coordination, … and 3) those designed to bring diplomatic/political benefits to Japan, including those aimed at regional stabilization and peace building (JICA 2003a: 8)." The effort has recently evolved into the economic partnership agreement approach of comprehensive alliances extending into the societal fabric, as opposed to US trade agreements limited to market relations. As Japanese negotiators made clear in the initial discussions of a Thai EPA, the goal was "not only a simple FTA but more comprehensive relations for the better future between the two countries (Miyagawa and Manawapatand 2003)."

The challenge

JETRO officials publicly lamented Japan's fading regional leadership already in 2004. A leading Japanese scholar went further recently, outlining specific areas of Japan's declining credibility in the region (Yoshizaki 2007). Inoguchi concluded that South Korea, China and Southeast Asian nations had prospered by accommodating foreign investment, in contrast to Japan's ruggedly autonomous model of development (2006). But not only is the Japanese precedent of insulated development called into question, Japan's commitment to the region has also been questioned. Her engagement with the world economy remains firmly rooted in the West, with the bulk of manufacturing FDI directed to the USA and EU (Lincoln 2005). Trade in technology remains firmly anchored in the USA. Despite the rise of regional economies, Japan continues to invest locally while retaining more extensive ties of capital and technology with advanced economies beyond the region. China's ominous development as a regional military power appears to have strengthened Japan's military alliance with the USA. Enthusiastic promotion of a regional "Arc of Values" (i.e., democracy, human rights, and market capitalism) with South Korea, Singapore, Australia, New Zealand, and India functions as a counter-weight to China's rising influence (Taniguchi 2007).

Despite the turmoil of transitions in the region and global markets, some things in Japan have not changed. The infusion of new capital from the USA and Europe into the NIEs, China and ASEAN, for instance, does not extend to Japan. Despite three decades as a major global trader and investor, the Japanese state has long discouraged foreign investment in its own economy. Even today, persisting low levels of inward investment appear to have scuttled plans to "further promote direct investment in Japan for the purpose of acquiring technologies and expertise from overseas (METI 2006a: 16)." The linking of investment with technology acquisition appears counter-productive. Anchordoguy found little change in the emphasis on indigenizing technology, or on government procurement in high-technology sectors (2005). Others look to more fundamental structural problems. Hatch cited persisting ties among elites in business and politics termed "relationalism," the core of a techno-industrial regime adept at incremental but not fundamental changes in industry standards (2003).

Transitions in Nikkei regional production chains provide a window on fundamental changes in Japan's position in the region over the past decade. Can firms adjust to regional opportunities for interactive learning by accommodating knowledge networks in the Japanese islands to wider regional resources? Adjustment would make possible a new leadership role in networks of innovation. Cooperation in innovation would promote development and cohesion in the East Asian Community. Commerce helped to drive Asian regional cooperation in an era of mass production. Can commerce now drive integration in an era of innovation?

3 Commerce and the East Asian community

Japan recovered from the Pacific War as the industrial leader in the region by the 1960s. Subsequent emergence of her East Asian neighbors three decades later as the "world factory" for electronics, textiles, and automobiles has radically changed the industrial landscape (Urata 2006). Japan remains a powerhouse in regional manufacturing as investor, innovator, and supplier of key components (METI 2006a). But the changing economic profile today of a region fueled by investment and trade by multiple nations has begun to dilute Japan's soft power in the region. Could a shift from assembly chains to knowledge networks propel Japan back into a leadership role in the region's economic community?

Trade and investment is the subject of this chapter. Trade flows reflect levels of technological expertise and extent of knowledge exchange. Shifts in the relative significance of major trade partners tell us of change in patterns and prospects for knowledge flows. Investment policies and practice offer a ground-level view of national innovation policy, and of emerging regional patterns of innovation (UNCTAD 2008a). Trends in Japan's regional patterns of trade and investment offer a portrait of her leadership role in the East Asian economy. Rather than simply upgrading production abroad due to structural adjustment demands at home, the data highlight the possibility of a new "knowledge nexus" under Japanese technological leadership in the region. Commerce can promote community.

East Asia

We do not find the familiar patterns of Japanese investment in the USA replicated in East Asia. Unlike forays into entertainment, real estate, as well as manufacture in the USA, Nikkei networks in East Asia have developed with a different scope, scale, and market. Affiliates of Japanese firms in Asia invest mainly in manufacture. Equally important is the large number of smaller Japanese firms moving abroad in the region, but not to the USA or Europe. Third, Nikkei networks in the region are more export-oriented than Japanese affiliates in the USA (Gaulier, *et al.* 2004).

Nearly five decades of investment from the 1960s has contributed to a ladder of technology and economic development today in sectoral investment and trade. Japan's vision for the region's GVCs was best conveyed recently as a "mountain-like structure" of supporting industries for supply of intermediate goods, supporting the industries of each nation in a "flexible manner (METI 2007b: 31)." Textile giants spearheaded Japan's initial regional push already in the late 1960s, soon joined by electronics and later automobiles. A ladder of trade and technology has emerged in Nikkei networks moving down from the four NIEs, to China, and then to four leading ASEAN nations.[1] South Korea and Thailand are Japan's leading manufacturing partners within the NIEs and ASEAN Four respectively. China is Japan's leading trade and investment partner in East Asia. Japan and the NIEs produce parts and intermediate goods, assembled in ASEAN and China (METI 2006a). For example, Nikkei affiliates in China and in ASEAN nations assemble cell phones and computers with computer chips imported from South Korea and Japan (Adams 2006).

Regional agreements

Extension of Nikkei trade and investment chains throughout Southeast Asia in the early 1990s spurred fears among her neighbors of a new "Yen Bloc (Frankel 1993)." Competition from below diluted Japan's leadership, as evident in South Korean trade and investment in China and in Southeast Asia, and now in China's rapid growth in trade and investment with Southeast Asia. But China and South Korea are only part of the new competition. Western investment in Southeast Asia grew rapidly after the Financial Crisis of 1997, particularly in electronics and automobile production. US industries also moved into South Korea following subsequent efforts at market liberalization and restructuring of banks and major manufacturers. American and European investors pushed ahead of Japan into China already in the early 1990s. In South Korea, Southeast Asia and China, investment grew in tandem with annual trade, linking production systems in a new nexus across borders in a region previously dominated by Japan.

Rapid change among her neighboring nations large and small has now brought those same anxieties back to Japan. China's emergence as the primary production center of the region frightens some (Sakakibara and Yamakawa 2004). Others see opportunity in a reorientation from a Japan-centered to a China-centered regional division of labor (Eichengreen and Hui 2006; Lee I. K. 2006). Dramatic growth in Japanese trade and investment with China since 2004 appears to have muted the debate for the moment on their neighbor as threat or opportunity.

At the same time, the Japan has reversed course on trade arrangements. Some state offices such as METI have dropped their earlier aversion to preferential trade agreements in favor of aggressive pursuit of bilateral and multilateral trade alliances. Economic Partnership Agreements have been concluded with Singapore and Thailand. Japan has made progress on her ASEAN + 1 Agreement, although still well behind China's implementation of their ASEAN + 1 agreement. Negotiations on a trade agreement with South Korea (ROK) remain stalled. Final ratification of the US–ROK Free Trade Agreement (FTA) by the respective legislatures would likely hasten Japan's efforts to at least initiate discussion with the USA on an FTA.

Trading partners

China's rise has not led to Japan's demise. Japan remains a central force in the "New Asia" with the region's largest economy and the highest standard of living. Japan's GDP of $4.3 trillion remains well ahead of China's, four times the size of South Korea's, and twenty times the size of Thailand's GDP (Table 3.1). Looking to relative prosperity, Japan's per capita GDP of $33,000 is close to double the size of South Korea's and far beyond the levels in China or Thailand. Third, Japan depends much less on foreign nations for technology, investment, or trade, than do her production partners. The ratio of inward FDI to GDP in 2005 highlights the prominent role of foreign capital and technology in South Korea (8 percent), China (14.3 percent) and Thailand (33.5 percent), but not in Japan (2.2 percent) (UNCTAD 2006a).

Japan's sustained prominence may surprise some, given the remarkable economic dynamism among her partner economies, as well as the prolonged recession in Japan herself. Annual GDP growth rates across fifteen years from 1990 of 9 percent or better in China, between 5 and 6 percent in South Korea, and about 5 percent in Thailand are evidence of remarkable economic expansion. Growth in per capita GDP in the brief six years from 2000 is even more striking. The figure has jumped from $945 to $2,000 in China, from $1,900 to $3,100 in Thailand, and from $10,800 to $18,300 in South Korea. Japan's per capita GDP in the same period dropped nearly $3,000 (IMF 2007). This statistical profile tells us much of macro-economic growth, but less of learning or technological capability. We learn more by looking to manufacturing, the largest contributor to overall R&D in most countries (Pilat 2006). Manufacturing for export also permits access to international standards and technologies not available locally, which offers a significant advantage in developing a national innovation system.

Some insist that Japan's networks of trade and investment in manufacturing have contributed to economic prosperity in the region through a "benign nexus of trade,

Table 3.1 Economic indicators, 2006 ($US millions)

Country/ Indicator	*Population (millions)*	*GDP at current prices*	*GDP/capita at current prices*	*Export value*	*Import value*	**Trade dependence 5-year average, 2001–05 (%) exports/imports*
Japan	127.7	4,302.09	33,667	595,269	516,202	11/9
China	1,307	3,248.52	2,797	726,327	660,222	27/25
South Korea	48.2	942.07	18,391	284,419	261,238	33/30
Thailand	65.1	239.87	3,575	109,848	118,112	58/57

Sources: APEC (2007), IMF (2007), KNSO (2006: 310–11), UNCTAD (2006a).

* Trade dependence computed as aggregate exports/nominal GDP of aggregate imports/GDP, and expressed as a percentage.

aid, and investment (Nakamura and Shinohara 2007)." But significant differences in knowledge flows among the three partner countries raise questions about local advantages in foreign trade with Japan. Trade[2] and investment[3] in electronics, automobiles, and textiles among Japan's leading Asian manufacturing partners is the topic of this chapter, the regional architecture of Nikkei production. In the next chapter I turn to innovation in the three production partners, tracking complementarity in the different national architectures of production.

Trade: China and South Korea

Three features distinguished Japanese production networks in electronics through the early 1990s: geographic concentration in Southeast Asia, asymmetric trade balances with East Asia, and hierarchical governance within relatively closed Nikkei networks (Ernst 2000). The major difference today is the China boom and the South Korean competition. Nikkei production networks in electronics, auto vehicles, and textiles now span the eastern seaboard of China as well as Southeast Asia. Japan's annual trade surpluses continue with Southeast Asia and South Korea, but not with China where Japan runs an annual trade deficit. But the latter deficit reflects a high volume of labor-intensive Chinese exports to Japan, rather than extensive technological development. South Korea tolerates annual trade deficits with Japan to import high-value-added components which can complement local manufactures in a profitable export trade. Thailand has emerged as Japan's automotive hub in Southeast Asia, but with "supporting" rather than competing local firms.

What might impede growth in trade and particularly in high technology exports from strengthening technology capacity in host countries? How can Asian hosts for Japanese firms anchor both manufacturing assets and Nikkei resources for product development? The character of the trade regime, and the structures of collaboration between home and host country deeply affect technology development. One problem is the processing trade, particularly in labor-intensive assembly industries such as electronics and apparel. Foreign firms may export the technology to the host country, assemble the product, and then export the product to a third country. This "processing trade" contributes to trade flows, but not necessarily to technological development (Lemoine and Ūnal-Kesenci 2004). Localization of production and fostering of complementarities in social systems of production can facilitate technological collaboration in Nikkei networks.

Region

Inter-industry trade in East Asia is particularly important in a region where a flow of intermediate goods fuel production chains for Japanese firms. A stream of products confined to networks of Nikkei firms represents a major share of Japanese export and import values. "Reverse imports," or the imports from global offshore Nikkei affiliates, accounted for nearly 17 percent of total value of imports to Japan in 2005, with imports from the Asian region alone amounting to 13 percent of the total value of imports. Exports to Japanese affiliates abroad amounted to nearly

34 percent of Japan's total export value for 2005. Exports to Japanese affiliates in Asia represented 13 percent of Japan's total annual export value in the same year (METI 2007c).

Looking beyond Japan to the region, we find trade playing a major role in Asian economic growth. A simple indicator such as total value of exports or imports divided by GDP provides one indicator of dependence on trade. The average annual rate for China from 2000 to 2005 was 27 percent for exports and 25 percent for imports. A smaller GDP coupled with massive external trade helps to explain higher dependency ratios for South Korea of 33 percent for exports and 30 percent for imports. In contrast, Japan's huge GDP moderates dependency despite growing annual values of trade, resulting in a trade dependency ratio of only 11 percent for exports and 9 percent for imports across the same five years (Table 3.1).

The USA, industrial Europe, China and South Korea have remained Japan's top export markets in that order from 2001 until China finally moved ahead of Europe in 2006 (JETRO 2006b; Yokota 2007). Incremental growth in the value of exports to the USA contrasts sharply with Japan's doubling of export values to South Korea, and tripling of the value of export to China in the same five-year period. Booming exports to South Korea and China cater to a growing demand for more value-added components in assembly operations. Japan's home country factories churn out ever greater volumes of critical components, as opposed to relocating production and perhaps upgrading the local industry in the host country. Smart states in South Korea and even China have pushed their own restructuring, which in turn fuels demand for more sophisticated products for local and export markets.

On the import side in same six years, the share of US products in Japan's trade fell by six percentage points to a 12 percent share in 2006. Imports from industrial Europe fell three points to an 11 percent share across the same period. Among Asian trading partners, import shares of the South Korean and Thai products held steady at 5 and 3 percent respectively, while imports from China increased by four percentage points between 2001 and 2006 to a leading 20 percent share (Table 3.2). Merchandise trade dominates Japan's economic transactions with her Asian neighbors, with machinery prominent in both exports and imports.

Trade: China

Trade and investment in Japan's closest geographical neighbors indicate growth with some diversification, but also technological development. The scale and relative importance of the China trade have improved dramatically in the past five years. Japan remains just behind the USA as China's largest export market, followed recently by South Korea. More than 20 percent of Japan's total imports in 2006 came from China, nearly twice the US share (Table 3.2). Looking back, the annual value of China imports has jumped from $37 billion in 2001 to $118 billion in 2006. The share of the three categories of electrical machinery (i.e., office machinery and computers, telecommunications, and other electronics) in total value of imports from China rose from 22 percent in 2001 to 31 percent in 2005, autos held steady at about 31.5 percent, and textiles and apparel fell from 30 to

Table 3.2 Japan trade, 2001–06 ($US billions, and percent of total Japan import or export)

Year/ Nation	*Total imports*	*Total exports*	*Import from China*	*Export to China*	*Import from South Korea*	*Export to South Korea*	*Import from Thailand*	*Export to Thailand*
2001	351 (100%)	405 (100%)	58 (16%)	31 (8%)	17 (5%)	25 (6%)	10 (3%)	11 (3%)
2002	336 (100%)	415 (100%)	61 (18%)	39 (10%)	15 (5%)	28 (7%)	10 (3%)	13 (3%)
2003	381 (100%)	469 (100%)	75 (20%)	57 (12%)	17 (5%)	34 (7%)	11 (3%)	15 (3%)
2004	454 (100%)	565 (100%)	94 (21%)	73 (13%)	22 (5%)	44 (8%)	14 (3%)	20 (4%)
2005	518 (100%)	598 (100%)	109 (21%)	80 (13%)	24 (5%)	46 (8%)	15 (3%)	22 (4%)
2006	579 (100%)	647 (100%)	118 (20%)	92 (14%)	27 (5%)	50 (8%)	16 (3%)	22 (4%)

Source: JETRO (2006b: Appendices 8–9).

20 percent (Sakamoto 2007). Information technology (IT) industries in electrical machinery are the major operators of value chains in East Asia (Adams 2006).

Electronics giants like Panasonic operate massive lines of production-sharing spread across Japan and China. Electrical machinery imports feed factory demand in Japan for insulated wire, parts for TVs and radios, and integrated circuits. Itochu has long been coordinating apparel imports from sub-contractors across China. Imports of textiles and electronics parallel Japan's sectoral investment in China. Japan's China trade and investment aligns as well with Japan's regional profile of textile and electronics trade across East Asia. Trade in motor vehicles, however, does not accurately represent the extent of Japanese region FDI in auto production (Urata 2006). Japan's auto firms move abroad to serve local markets, or to export to a third country. Trade figures do not adequately capture the China commitments of giants like Toyota or Denso.

A new partnership?

Demand for Chinese imports has grown in tandem with Chinese demand for Japanese higher value-added manufactures. JETRO Chairman Yasuo Hayashi attributed the recent increase of Japan's exports of parts and materials to Nikkei plants operating in China (2007). A METI survey of Japanese manufacturing affiliates reported sales in China for 2007 of $95 billion. The total included $20.4 billion in exports back to Japan, and an additional $49.7 billion of domestic sales in China (METI 2007a).

Japan's exports to China of only $31 billion in 2001 tripled to more than $92 billion in 2006. Exports of electric equipment alone jumped from $8.1 billion in 2001 to \$20.7 billion in 2006, but still held the same 26 percent share of the export total as in 2001 (JETRO 2006b: Appendix 6.11). With little Chinese investment in Japan's textile industry, and weak demand in China for specialty fibers, Japan's annual textile exports to China remained constant between 2001 and 2005 at about \$3 billion or 4.5 percent of total exports to China. In the same period, transportation equipment including exports of vehicles and auto parts to China jumped from \$1.3 billion to \$4 billion or about 5 percent of total exports in 2005 (JETRO 2006b: Tables 4.3, 34).

Japan ranks atop the world's leading exporters to China, followed by industrial Europe. South Korea ranks third, having moved ahead of the USA already from 2002, and may soon surpass Europe in the China market as well. Japanese firms lost ground to Korean firms despite holding on to the top position in the China trade. Japan's share of China's total annual imports declined from 17.5 percent in 2001 to 14.5 percent in 2006, whereas Korea's share grew from 9.5 percent to 11.5 percent. Looking to the other side of the trade ledger, Japan's exports could not keep pace with Japan's appetite for Chinese imports. Growing imports from China resulted in an annual bilateral trade deficit for Japan, reaching \$26 billion in 2006. The deficit reflects in part the success of Japanese investors. JETRO Chairman Hayashi observed recently: "The majority of Japan's imports from China are finished products made by Japanese firms operating locally, such as business equipment and audio-visual devices (Fung and Iizaka 2004; Hayashi 2007)."

Despite the growing value of Japan's imports from China, the rapidly expanding value of China's global exports has diluted Japan's comparative importance for Chinese exporters. Japan's portion of the total value of China exports fell from 17 percent in 2001 to only 10 percent in 2006 (GTIS 2007). But other leading trade partners such as the USA (21 percent), industrial Europe (18 percent), Hong Kong (16 percent) and South Korea (4.6 percent) largely maintained their shares of China exports over the same period. In sum, China's importance as a trade partner to Japan increased markedly over 2001–06, both as an export destination and a source of imports. Japan's relative significance as a trade partner for China, however, faded in the same period.

Trade: South Korea

In contrast to trade deficits with China, Japan enjoys an annual surplus with South Korea, her fourth largest import partner. Imports from South Korea jumped from \$17 billion in 2001 to \$27 billion in 2006 on an annual basis (JETRO 2006b: Appendix 6.14). Manufactures accounted for 77 percent of Japan's annual imports from South Korea between 2001 and 2005. Among these, the three categories of electrical machinery accounted for a steady 21 percent of total value of Japanese imports from South Korea in 2001 and 2005. How important is the sector for the South Korean economy? The value of annual production of electronic merchandise in South Korea increased from \$53.6 billion in 2001 to \$93.5 billion in 2004, equal to 13 percent of the nation's GDP (KNSO 2006).

Electronics contribute mightily to South Korea's expanding global trade, spurred in part by the China factor. China's emergence in assembly of electronic products highlights the changing nexus of trade and investment in the region. Competition with China is forcing South Korea to diversify export markets, and to upgrade production as Japan relocates production lines from South Korea to China. China's share of Japan's electric machinery imports rose from 21 percent in 2001 to 31 percent in 2005 alone (Sakamoto 2007). But Japan imports different products from the two nations, reflecting difference in the technology ladder among trade partners. Integrated circuits account for better than half of Japan's imports of electronics from South Korea, but only 6 percent of the imports of electronic machinery coming from China.

A fading partnership?

Appetites for South Korean products in the USA and Japan have played a part in Korea's ascendance as a major global trading nation. Refined fuels from South Korea's extensive petrochemical complex, and steel from its huge steel works accounted for about 15 percent of imports from Korea in 2005, but the broader category of machinery manufactures, including electronics, accounted for 61 percent of all exports that year (MOCIE 2008). A flow of South Korean chemical products captured 8 percent of the total Japanese imports from the peninsula by 2005, and iron and steel imports increased from a 5 percent share in 2001 to 9 percent in 2005. At the same time, Japan's imports of South Korean textiles dropped precipitously from nearly $1 billion in 2001 to $734 million in 2005 just as imports from China boomed.

Japan's relative status as a destination for South Korean products has faded despite incremental annual growth in value, due to the exponential growth in Korea's exports to China. Exports destined for Japan faded from 11 percent of South Korea's total exports in 2001 to only 8 percent in 2005. Other trading partners saw their market share decline as well. The share of South Korean exports destined for the USA fell from 21 percent to 15 percent, even as the share of Korean exports destined for Europe held steady at 14 percent annually. Meanwhile the share of South Korean exports destined for China jumped from 12 percent to 21 percent over the same period. In terms of the share of single country destination in total export value, Korea has reoriented itself to the China market at the expense of the USA and Japan, despite incremental annual growth of trade values in the latter markets as well.

South Korea's growing appetite for Japanese products, on the other hand, merits closer attention. South Korea's import bill from Japan of $25 billion in 2001 doubled to a total value of $50 billion in 2006. Components for electrical machinery, general machinery, and iron and steel alone account for better than half the total annual value. But despite growth in annual value, South Korea's total share of Japanese exports in the three categories of electrical machinery declined from 29 percent in 2001 to 21 percent in 2005, with component exports probably redirected to production networks in China. Meanwhile, South Korea expanded its share of total Japanese exports in other sectors. For instance, her share of Japanese total exports of general machinery increased from 16 percent to 18 percent in the same period. Japan's exports of iron and steel to South Korea likewise increased from 9 percent of total annual trade value to 12 percent by 2005, with similar growth in exports of chemicals from 15 percent to 17 percent of total trade value in 2005. Electrical components and specialty iron and steel, and chemical products, provide the more sophisticated materials necessary for production of South Korea's own exports.

What might explain South Korea's growing demand for general machinery from Japan, doubling in value from $4.3 billion in 2000 to $8.4 billion in 2005? Reliance on Japan for key components and high technology appears to grow in tandem with the expanding volumes and growing sophistication of South Korea's global exports (Mukoyama 2005). Domestic production in South Korea continues to draw on Japanese parts, intermediate goods and equipment that cannot be procured domestically (Wong, *et al.* 2004).

A new bipolar nexus?

China looms large in the changing format of South Japan–Korea trade and production ties. Incremental annual growth in South Korea's imports from Japan has not kept pace with the overall expansion of import demand in Korea. Japan's share of the total South Korean import market fell from 19 percent to only 17 percent between 2001 and 2006, just as China's share of the same market increased from 9 percent to 16 percent (Kameyama 2007). The "China turn" on the peninsula represents a benign nexus of trade and investment of its own, with South Korea ranking now just behind Japan as China's leading investor (Lee I. K. 2006).[4] South Korea's trading position in Japan's regional production network has changed: Japan has faded as an export destination, and now must compete with China as South Korea's leading source of imports.

Quite apart from China, slow growth in exports to Japan coupled with a growing appetite for Japanese imports has resulted in a trade imbalance that will be difficult to sustain. Japan reported $25 billion of exports to South Korea in 2001 and $17 billion of imports. Five years later, Japan registered $50 billion of exports but only $27 billion of imports. Persisting imbalances in bilateral trade reflecting a division of labor in complementary industries such as electronics can often be tolerated if offset by exports to other markets. Trade imbalances within a regional production network of competing Japanese and Chinese centers may well erode the benefits for South Korea of the Japan-led nexus of trade and investment. The contrast with South Korea's annual surplus in the China trade across the same years can only aggravate the problem and further dilute Japan's soft power.

Investment: China and South Korea

The past two decades have brought growing volumes of foreign investment to China, South Korea, and Thailand (Table 3.3). Asia has captured a growing annual share of Japan's outward investment flows (UNCTAD 2008b) in the past few years, although North America still anchored 40 percent of Japan's direct investment assets through 2005. Japanese firms registered a 38 percent share of total inward FDI to the USA, a smaller 24 percent share in the fifteen countries of the EU, and a still smaller 22 percent share in Asia (Table 3.4).[5] What is perhaps most surprising is Japan's comparatively slow penetration of the Chinese market, evident in Japan's 6 percent share of China's total inward FDI stocks. Nikkei investment has not kept pace with the growing Japan–China trade.

Turning from regions to Japan's FDI stock in Asia, a familiar ladder of investment emerges in 2005. The Asian share alone of global Japanese FDI stocks amounts to $87.3 trillion, distributed among the four NIEs (37 percent), the ASEAN 4 (31 percent), and China (28 percent). Among Japan's production partners, South Korea ranks second only to Singapore among FDI destinations in the Asian NIEs, although Korea is a far larger Japanese trade partner. South Korea also produces far larger annual values of electronics, autos, and textiles, where it competes with Japan. Among Nikkei production partners in ASEAN, Thailand's 3 percent share of Japan's global direct investment assets far surpasses that of any other member.

FDI "stocks" (UNCTAD 2008c) offer only one picture of investment, capturing the historical trajectory but not current trends. Fluctuating annual investment totals suggest changes in investment profiles (Table 3.5). Annual Japanese FDI

Table 3.3 FDI stocks in China, South Korea and Thailand, 1980–2005 ($US billions)

Year/Country	*China (% GDP)*	*South Korea (% GDP)*	*Thailand (% GDP)*
1980	28	1.3	0.98
1985	40.5	2.1	1.9
1990	84 (8.4%)	6.6 (2%)	8.2
1995	101	9.4	17.6
2000	193.3 (17.9%)	37.4 (7.3%)	29.9 (24.4%)
2005	317.8 (14.3%)	63.1 (8%)	56.5 (33.5%)
2000 Japan share	8.6 (4%)	4.19 (11%)	4.76 (16%)
2005 Japan share	24.4 (7%)	8.17 (12%)	11.5 (20%)

Sources: UNCTAD (2008a). Percentages of gross fixed capital formation provided by UNCTAD (2006b). Japan share drawn from JETRO (2008).

Table 3.4 Japan's direct investment assets, end of 2005

FDI/Destination	*World*	*Asia*	*Western Europe*	*USA*	*PRC*
Amount ($US trillions)	*384.4*	*87*	*91*	*148*	*24*
% of total outward FDI	100%	22%	23%	38%	6%

Source: (JETRO 2008).

Table 3.5 Japan's global investment flows, 2005–06

FDI/Destination	*World*	*Asia*	*Western Europe*	*USA*	*PRC*
Amount 2005 ($US millions)	*45,461*	*16,188*	*7,509*	*12,126*	*6,575*
% of total outward FDI	100%	35%	16%	26%	14%
Amount 2006 ($US millions)	*50,165*	*17,167*	*18,029*	*9,297*	*6,169*
% of total outward FDI	100%	34%	35%	18%	12%

Source: JETRO (2007c).

flows to China in the past two years reflect better than double their ratio of FDI stocks. Indeed, Japan's annual FDI flows to China from 2001 to 2005 averaged $4.5 billion, a rise of 300 percent from FDI flows in the previous five-year period. On the other hand, declining annual flows to the USA in the past two years indicate a considerable drop from their overall ratio of FDI stocks for 2005. Nor can we overlook the prominence of the Asian region in annual flows, far higher than their ratio of FDI stocks. Trends evident in the data include: a) the continuing but somewhat diminished prominence of the USA as a destination for FDI flows, and b) the growing importance of the Asian region and especially of China. Finally, we find c) a sectoral priority on manufacturing in Japanese global FDI, but particularly in FDI to the Asian region. Manufacturing investments accounted for 79 percent of Japanese FDI to Thailand in 2006, 78 percent of FDI to China, but only 59 percent of Japanese FDI to South Korea (BOJ 2007a).

Investment: China

Rising production costs at home and trade frictions abroad drive Japanese firms offshore. But such "push" factors play a role only in tandem with "pull factors" in target nations, particularly local state policy in East Asia. Japan, South Korea, and China represent three quite discrepant approaches to inward FDI in technology development. China opted to welcome inward FDI from the outset to gain foreign technology. Foreign investment flows turned to China soon after implementation of Deng Xiaoping's "Open Door" policy from 1972. The nation reported a jump in inward FDI stock from $28 billion already in 1980 to $40 billion five years later, $84 billion by 1990, and finally $101 billion by 1995. The amount nearly doubled then to $193 billion by 2000 (Table 3.3). Foreign investment and technology has played a major role in China's economic growth, with FDI stocks from abroad in 2005 amounting to 14.3 percent of China's total GDP, as compared to only 5.4 percent of GDP in 1990. China's ratio of FDI to GDP is seven times larger than the comparable figure for Japan and quite similar to the figure for the

USA (13 percent). But as with South Korea, what is remarkable is the pace of FDI growth as well as the scale.

Pace of FDI growth

Clearly Japan has not set the pace in China, despite the extent of overseas development assistance (ODA), and the early market penetration of firms such as Itochu and Panasonic. Chinese constraints on FDI apparently discouraged Japanese investors, particularly in the auto industry. The Chinese state imposed multiple controls to insure technology transfer, the participation of state-owned enterprises, and some dispersion of industrial investment beyond the eastern seaboard. Improvements in transportation and energy supply recently have eased investors' concerns, but complex and sometimes inconsistent legal procedures continue to constrain foreign investment. For instance, the Japanese business community in China regularly complained of constraints on finance until recently, and continues to fret about inconsistencies in protection of intellectual property rights. In any case, the trajectory of Japanese investment in China from 1980 does not align with the rosy picture of growth cited above for global FDI to China.

Japanese FDI to China fluctuated around $500 million annually from 1988 to 1992, but then jumped dramatically from 1995 (Table 3.6). Chinese state incentives for foreign investors perhaps account for the expansion of annual flows to the level of $210 million then from 1991 to 1995 (Fung and Iizaka 2004). China absorbed only 19 percent of total Japanese FDI to Asia in the five years 1996–2000, but then 44 percent in the subsequent five years through 2005.

New preferential policies for foreign firms, and China's entry into the World Trade Organization in 2001 helped to raise annual investment flows to an average of $450 million from 2001 to 2005. Japan's share of FDI stocks in China nearly doubled from 4 percent to 7 percent in the same period. Japan's Ministry of Economy, Trade and Industry (METI) estimated employment at Japanese affiliates in China at over one million by 2004 (Sakamoto 2007). Most recently, we find Japanese investment pouring into China with an annual average flow of $6.2 billion between 2004 and 2006 (JETRO 2008). Given only incremental growth at best across two decades of Japan's FDI flows to China, the recent jump in investment

Table 3.6 Japanese FDI flows to East Asia, 1996–2005 ($US millions)

Recipient/Period	*1996–2000*	*Percent of total Japanese FDI flow to Asia*	*2001–05*	*Percent of Japanese FDI to Asia*
China	6,774	19%	21,265	44%
South Korea	1,728	4.9%	3,927	8%
Thailand	5,984	17%	6,792	14%
Total to Asia	*34,620*	*100%*	*47,721*	*100%*

Source: FDI flow based on international investment position JETRO (2008).

is a radical departure from the pace and scale of Japan's manufacturing commitments in China. The changes highlight a reorientation of Japan's nexus of trade and investment in East Asia.

The electrical machinery sector has consistently led Japanese foreign investment to China. The textile industry often drew the largest number of ventures on an annual basis through 2000, but usually with comparatively small capital investments (Fung and Iizaka 2004). Electronics has consistently dominated investment, supplemented by numerous textile operations, but the transport sector began to draw investment in the 1990s, and then expanded dramatically in the past few years. Affiliates in electric machinery reported employment of 476,000 in 2004, with a further 120,000 workers in transport machinery, and 78,000 in textiles and apparel. The investment patterns are reflected in strong trade in textiles and electrical machinery with China. As with South Korea, the transport machinery sector investment is not reflected in bilateral trade. But whereas the South Korean auto industry is a major exporter, the Chinese auto industry continues to grow mainly in the domestic market.

Investment: South Korea

South Korea's early preference for indirect rather than direct investment resulted in a pattern of licensing and loans, maintaining relative isolation from inward FDI. This paralleled Japanese state discouragement of inward FDI in the development process, persisting today in the imbalance between Japanese outward FDI stocks of 8.5 percent of GDP, as compared to inward FDI stocks of only 2.2 percent of GDP. But Japan enjoyed relative autonomy from the turn of the nineteenth century in gaining foreign technology without equity shares. South Korea developing much later had little choice but to welcome foreign investment for the initial establishment of major industries. If China has welcomed foreign investment in manufacture, and Japan largely excluded direct foreign investment in her initial decades of post-World War II industrialization, South Korea has carved out a path somewhere between engagement and exclusion.

Joint-ventures helped to pioneer South Korea's rapid industrial development of export industries from the 1960s. Japan's leading synthetics firms joined with local firms like Sunkyong and Kolon for initial development of the industry (McNamara 2002). General Motors of the USA and Mitsubishi later joined with Daewoo Motors and Hyundai in the initial stages of automobile production (McNamara 2005). Japanese and US firms also provided much of the early technology for the electronics firms. Grants and soft loans from Japan of $800 million in the Japan–Korea Treaty of 1965 opened a new flow of Japanese technology into state-targeted industries. Apart from the major joint-ventures, Japanese SMEs moved quickly into the special export zones, supported by South Korean state incentives of tax reductions and access to land on industrial estates (Han 1994). Subsequent state policies to indigenize technology and localize production brought an end to most of the joint-ventures, although Japanese SMEs remained.

Strong state

A strong South Korean state under military leaders like Park Chunghee and then Chun Doo-huan through 1987 tolerated investment from abroad only for acquisition of critical technology in infant industries. South Korea reported FDI stocks of only $1.3 billion in 1980 (Table 3.3). Limited market opening subsequently permitted expansion of FDI stocks to $6.6 billion a decade later, an expansion of close to 600 percent but off a modest base. The total amount reached $9.4 billion in 1995. Insulation strategies faded by the late 1980s and early 1990s as South Korean export industries reached a technological ceiling with foreign firms reluctant to provide technology without equity stakes. Both pull factors of foreign pressure to liberalize, and push factors from globalizing South Korean industries finally brought change.

Liberalization of FDI flows after the 1997 Financial Crisis brought a new wave of Japanese investment, in tandem with US investment. Annual inward FDI flows averaged only $1 billion from 1991 to 1995, but then jumped to $5.8 billion across the next five-year period (BOJ 2006). South Korea's FDI stocks grew better than 400 percent through 2000 to a level of $37 billion. FDI stocks in South Korea nearly doubled between 2000 and 2005, growing from $37 billion to $63 billion (UNCTAD 2008a). FDI stocks from abroad in 2005 amounted to 8 percent of total GDP, nearly four times the comparable figure for Japan, but still small compared to FDI leaders like Britain (37 percent) or the European Union (33 percent).

More than scale of investment or investment partners, it is the rapid pace of inward investment that sets South Korea apart from its neighbors. Inward FDI stocks grew from 2 percent of GDP only fifteen years ago, to 8 percent in 2005 – remarkable ratios given the growth in the GDP during the same period. South Korea's $10.8 billion of Japanese FDI assets hardly compares with Nikkei assets in Thailand ($15 billion) or China ($30.8 billion) (BOJ 2007a).[6] Moreover, 40 percent of Japan's FDI assets in South Korea represent reinvested earnings, not equity capital (BOJ 2007b). Four decades of change in inward FDI policy tell us much of Korean state, society, and innovation strategy.

One can hardly ignore the prominence of the South Korean state in sectoral policy and practice in South Korea's technological development. Techno-nationalism identified indigenous technology as a key to national strength, crucial in the struggle for survival on the peninsula with the hostile regime of North Korea. The South Korean state based its legitimacy on economic development and localization of foreign industrial resources. Inward FDI policy also reflects South Korea's close interdependence with foreign sources of technology, particularly from Japan and the USA. As one scholar recently concluded regarding South Korea's technology acquisition process, "the role of foreign MNEs has been very significant for Korea (Cyhn 2001: 41)." Finally, the role of trade, investment, and exporting, often manufacturing as sub-contractor for larger firms, was likewise formative in South Korea's process of technological learning.

Nikkei investment

Employment at Japanese affiliates in South Korea highlights the curious trajectory of Japanese investment. Data from databases of the Toyo Keizai profile a height of 176,000 workers in 1995 before the Financial Crisis, with 14,000 in electric machinery, 9,000 in textiles and apparel, and 8,000 in the transport machinery sector. Affiliates registered nearly a 50 percent drop in total employment in 1998 to only 75,000 workers. Employment continued at that level through 2005, with 11,500 in electric machinery, 7,600 in transport machinery and a similar number in chemicals, and still 5,200 in textiles and apparel (Kameyama 2007; UNIDO 2008). Employment at Japanese affiliates reflects the concentration of Japanese investment in manufacture rather than services, but remains a very small portion of South Korea's total labor force of some 24 million.

Japan's 12 percent share of South Korea's total inward stocks ranked just behind those of the USA among Korea's top FDI partners. Manufacturing alone captured nearly three-quarters of Japan's total $8.1 billion investment in the peninsula in 2005. Most of Japanese investment over the past two decades has flowed into South Korean chemicals, transportation equipment, and electrical machinery. The South Korean state secured foreign investment for initial development of infrastructure in chemicals, automobiles, and electronics. Some industries in turn became suppliers for Japan, evident today in South Korea's exports of refined chemicals to Japan.

A similar process is evident in electrical machinery with initial investments from Japan, but now an independent industry in South Korea exporting finished products to Japan. Complementarities in machinery sectors, chemical products and textiles, as well as the prominence of electrical machinery, distinguish the export flows of both countries (Urata and Abe 2005). But in the transportation industry, investments of Japanese SMEs in the parts industry contribute to assembly and eventual export to third countries. Transportation equipment is a major export commodity of South Korea to countries other than Japan.

Transition from FDI destination to FDI investor may be the most intriguing change in Japan–South Korea ties. Outward FDI has risen from less than 1 percent of South Korean GDP in 1990, to 4.6 percent in 2005. Rather than simply scale, the real significance of the change lies in the directions of investment. China attracted 40 percent of South Korea's realized outflows of FDI between 1992 and 2005. At best the world's eleventh largest economy, South Korea provided nearly 9 percent of China's inward FDI in 2005. China has not shown a similar investment interest in South Korea, precluding a stronger manufacturing linkage between the countries for the time being.

Not only does China provide a counter-weight to Japan's leadership of the trade/investment nexus, but trade and investment imbalances in the South Korea–Japan nexus have diluted Korea's link to the Japan center in favor of China. Reviewing investment flows between Japan and South Korea in the decade from 1995, one scholar estimated Korean outward FDI to Japan at only 10 percent of the total Japan directed to Korea (Kameyama 2007). South Korea has invested in China in preference to Japan. Investment profiles in South Korea and China reflect the new division of labor in the Asian region with China joining Japan as a major center.

Trade and investment: Thailand

Thailand is Japan's leading manufacturing partner among the ASEAN nations. Still largely an agrarian nation, the agriculture sector employs 38 percent of the labor force but produces only 10 percent of GDP (ESCAP 2007: 152; Ramstatetter 2007). Thailand has not been able to develop a "full-set" industrial structure comparable to its three much larger neighbors to the north. Instead, Thailand hosts major manufacturing MNEs from Japan, the West, as well as from South Korea and China. A structural pattern has emerged of larger foreign firms working with a smaller base of local "supporting industries," mainly SMEs.

One result of the imbalance between extensive trade and a relatively small GDP was the average five-year trade dependence ratio of 58 percent in exports and 57 percent in imports through 2005 (Table 3.1). Similar to its northern neighbors China and South Korea, the Thai economy over the period 2001–05 shows growth and diversification, with a reorientation to Chinese markets and investment. Trade liberalization and integration among the ASEAN nations may eventually permit new economies of scale in textiles, electronics, and especially autos. But the problem remains of trade and investment without comparable local development of technology.

Trade: Thailand

The value of Thailand's merchandise exports jumped from $65 billion in 2001 to $110 billion by 2005 (Ramstatetter 2007). Among leading products across five years, the share of transport equipment doubled from 4 percent to 8.5 percent. With ambitions to make Thailand the "Detroit of Asia," Japanese and Western automakers increased production from 585,000 units in 2002 to 1.1 million units in 2005 (KNSO 2006). Electrical machinery held steady at about 29 percent while textiles and apparel fell marginally from 8 to 6 percent. Leading destinations for Thai products include the USA, where the share of total Thai exports declined from 20 percent in 2001 to 15 percent in 2005, and Japan where the share of Thai exports faded from 15 percent to 13 percent. South Korea's share of Thai exports held steady at about 2 percent, but China's share doubled from 4 to 8 percent. Although some of the exports to Hong Kong may be included already in the China share, the Hong Kong portion in Thai exports of 5 percent would raise the wider China share in Thai export trade of 2005 to a level similar to that of Japan.

Electronics and agricultural goods rank among Thailand's leading exports to Japan. The three categories of electric machinery declined from a combined share of 30 percent of export value in 2001 to 25 percent in 2005. Agricultural exports declined from 22 percent share to a 15 percent share in the same years. In contrast, the portion of general machinery in annual export value rose from 39 percent to 43 percent in 2005, despite decline in exports of electric machinery. Curiously, exports of transport machinery held steady at about 2.5 percent of total value of exports, despite the doubling of local production. An extensive base of Japanese auto assembly and component makers in Thailand produces mainly for the local market. Thai textile and apparel exports declined marginally from 4 percent in 2001 to 3 percent in 2005, while the share of chemical manufactures remained constant across the period at about 5 percent. Consistency or marginal change in shares of total export values which nearly doubled in the five-year period suggests vibrant industrial sectors, but tells us little of indigenous innovation. The major decline in export share was registered in imports of machinery, declining from 45 to 38 percent across five years. Within this sector, the export value share of electrical machinery fell from 28 percent to only 22 percent in 2005.

Turning from exports to imports, we find that Thailand's total value of merchandise imports nearly doubled from 2001 to a sum of $118 billion by 2005. Japan's 22 percent share of Thai imports held steady through 2005, far and away the leading source of imports. The US share faded from 12 to only 7 percent by 2005, and the share of industrial Europe in the Thai import market declined from 14 to 10 percent. Meanwhile China increased its share of the Thai import market from 6 to 9 percent. Thailand has not emulated South Korea's turn to the China market. The nation has maintained closer trade ties with Japan than has South Korea, although both Thailand and Korea suffer a trade imbalance with Japan. Extensive Japanese investment in the Thai manufacturing base, and consequently a major role in Thai exports, may partially explain the comparative resilience of Japanese ties. Japan is also the major source of official financial flows from abroad, with the Japan Bank for International Co-operation (JBIC) providing critical funds for major infrastructure projects such as the New Bangkok International Airport (Ramstatetter 2007).

The nation reported big import gains in basic manufacturing inputs such as mineral fuels, jumping from 12 to 18 percent of export values in 2005, and iron and steel which grew from a share of 4 percent in 2001 to a 7 percent share of export totals by 2005. Imports of chemical manufactures declined marginally to a 10 percent share in 2005, as did the textile and apparel portion to a 2 percent share in 2005. Despite huge growth in auto production during the period, imports of transport machinery held steady at about 5 percent.

Japan partnership

China's growing role in Thai trade is widely recognized. China stands now as the second largest source of imports, and fourth largest export destination. Moreover, China has a crucial role, not only in bilateral trade with Thailand "but also in the fierce competition between the two countries in third markets (TN 2007b)." Yet persisting dependence on imports from Japan, and Japan's continuing status as a leading export destination maintain the trade directions in this nexus of trade and investment. The prominence of manufactures in the bilateral trade with Japan further reinforces the nexus.

Curiously, neither trade liberalization within ASEAN, nor the rise of China has seriously eroded the Japanese share of Thai trade. Indeed among sources of imports, only Malaysia increased its share of the Thai market. On the export side, Thailand did increase the share of products destined for Indonesia and Malaysia, but this might be explained by transfers within Japanese production networks in the region. The intensity of Thai trade dependency on both the export and import sides lend the Japan link even greater weight in the Thai economy.

Investment: Thailand

Thailand has generally welcomed foreign investment, but in the absence of a strong local technology policy, has not gained much technological advantage. National content rules and local partnership initially fostered local supporting industries in metals, tool and die firms in the auto industry, and electrical components (Felker 2004). Unlike South Korea's supporting industries in the 1970s, however, the Thai industry emerged as a "contingent cluster" of mainly Japanese production chains, rather than an indigenously owned national export industry.

Liberalization efforts have not been balanced with an effective national innovation system. The "Brand-to-Brand Complementation Scheme" was approved in 1989 to permit tariff-free movement of components for auto assemblers among leading ASEAN nations. The arrangement permitted tariff reduction or exemption for movement of parts for assembly across ASEAN countries. Toyota Motor Company was moving products among production sites in Thailand, Malaysia, the Philippines and Indonesia by 1993 (MERI 2006b; Yoshimatsu 2002b). The Financial Crisis of 1997 accelerated regional liberalization in the ASEAN Industrial Cooperation (AICO) Agreement, including other industries such as auto parts makers and electronics firms in the cross-border supply effort (Sasuga 2004).

Foreign investment in Thailand began to flourish from 1985, moving up from nearly $2 billion in 1985 to $8 billion in 1990, and then more than doubling to $17.6 billion by 2000, but still representing only less than 10 percent of GDP (Table 3.3). Direct investment totaling $56.5 billion in 2005 boosted the dependency ratio from 9.7 percent to 33 percent. The figures contrast with far lower ratios in South Korea (8 percent) and China (14 percent). This indicates considerable dependence in Thailand on anchoring international investment in local electronics, auto, chemical, and textile production. Aggravating the situation is the fact that few Thai firms command the resources to move abroad: total outward FDI stocks represent only 2.3 percent of total Thai GDP.

Japanese investment

Japanese investors flocked to Thailand following the appreciation of the yen after the Plaza Accords of 1985. The textile sector usually led in the annual number of projects through 1995, but with relatively small capital investment compared to much larger projects in electrical machinery (MOF-Japan 2007). The capital flow from Japan to Thailand peaked in the years immediately before the Financial Crisis, amounting to an annual average of $1.2 billion from 1996 to 2000, triple the annual average to South Korea, and just behind the annual average to China (Table 3.6). Investment cooled more recently, with the annual investment totals through 2005 reaching on average only $900 million, comparable to levels of Japanese annual flows to South Korea, but now only one-quarter of investment flows to China.

Annual investment flows to Thailand still have not returned to the pre-Financial Crisis levels, despite steady FDI growth in the transport and electric machinery sectors. Japanese firms have begun to hedge their bets with a "China + 1" strategy of diversifying investment into Thailand and Vietnam (Hosokawa 2007), and Japanese firms have increased their share of Thailand's total inward FDI stock from 16 percent to 20 percent through 2005. This prominent share of private capital investment, coupled with extensive official finance flows from JBIC, would indicate a far more important role for Japan in the Thai economy than we find in either the South Korean or Chinese economies (MOFA 2006).

Leading Thai industrial sectors include consumer electronics, autos, and chemicals, as well as a smaller sector of textiles and apparel. The value of annual production of electronic merchandise in Thailand increased from $14 billion in 2001 to $21 billion in 2004, representing 13 percent of the nation's GDP that year. Transport machinery is also a major contributor to the economy. The nation almost doubled vehicle production from 585,000 units in 2002 to 1.1 million in 2005 (KNSO 2006). Japanese firms play a central role in each of these sectors. Estimates based on corporate reports set employment at Japanese affiliates in Thailand in 2005 at about 414,000, or about 7 percent of the 5.5 million Thai workers employed in manufacturing (Ramstatetter 2007). Among leading employers, affiliates in electric machinery alone accounted for about one-third of total employment with Japanese affiliates, with a further 18 percent of Japanese affiliate employees in transport machinery. Combined employment in chemicals, textiles, and apparel accounted for 8 percent of employment at Japan's affiliates in Thailand.

Conclusion: a new knowledge nexus?

Japan's ODA has also played a role in the benign nexus of trade and investment. With gross ODA assistance totaling $1.6 billion, China ranks second only to India among leading aid recipients. Thailand ranks fourth with a total of $765 million (Manning 2007). The flow continues, with annual aid of $750 million on average to Northeast Asia between 2000 and 2004 (Nakamura and Shinohara 2007). A redirection of funds from infrastructure projects such as roads, bridges, and power plants, to technical cooperation and policy planning is one change apparent among production partners. My question is whether the organization and technical side of private investment in manufacturing can prosper with a similar emphasis.

New partners have significantly altered the earlier integration of Nikkei trade and investment in Asian production chains. China has emerged as a competing center in trade and production, though not as yet in direct investment abroad. South Korea has diversified its trade ties, welcomed a variety of foreign investment sources, and moved aggressively to expand its outward FDI, and that without much investment in Japan. Thailand has also seen China rise as a major trade partner, though

Japan retains a more prominent role in the Thai economy than in either China or South Korea.

America's declining status in regional trade and investment also affects Japan's leadership role in the regional division of labor. The slower pace of growth in US trade and investment has eroded the comparative prominence of the US economy in the region. Japan and the USA are bound not only by military and economic ties, but also by a more fundamental agreement on contract and law as a basis for economic exchange. Beyond bilateral ties, the sustained partnership and cooperation of the USA and Japan in multilateral fora such as the OECD, World Bank and IMF have promoted common industrial standards in Asia's production networks. Working often in tandem, the two nations have contributed to enforcement of intellectual property rights critical for effective knowledge transfer (INT 2007c). US consumers have also played a major role as end-users for many of the products in Japan's production networks. The success of South Korea and Thailand in registering trade surpluses with the USA provides a balance to trade deficits with Japan. A declining role for US exports to the region, and for US investment in the region, may prompt more serious efforts to curtail those surpluses.

Technology

Technology is one of the forces driving the change in the nexus of trade and investment. Japan confronts a new technological competition with major South Korean *jaebeol* in regional markets of electronics, autos, and synthetic fibers. A JETRO report recently lauded the successful penetration of the China market by Hyundai Motor and LG Electronics (JETRO 2006a). Perhaps more unsettling is the fact that domestic electronics firms in China have proven tough competitors for Japanese firms, and will soon extend abroad. Haier Electronics Company has already joined with Distar Electric Corporation Public Company in Thailand for marketing and production, and with Japan's Sanyo Electric Company for production of refrigerators in Thailand. Shanghai Automotive Company Ltd purchased the South Korean carmaker Ssangyong in 2004, and 10 percent of the shares of Daewoo Motor Sales Corporation (METI 2005). Auto parts firms and assemblers in China have only begun to test international markets. Japan faces a far less intense competition in Thailand for local engineers and researchers as US and European firms in electronics and autos localize production.

Technology policy takes center stage in the official industrial strategy supported by Japanese government agencies. A JETRO White Paper on trade and investment recommended market alliances with East Asian corporations, and business alliances with global firms for product development (JETRO 2006a). What drives this change is the fact that annual growth in overseas income has continued to outpace growth in GDP since 2003. JETRO sees economic expansion in other East Asian economies as a means to leverage Japan's own growth. Market alliances would permit rapid response to fast-growing Asian markets.

One might conclude that local firms closer to the market are better positioned to learn and respond. Beyond marketing, alliances for product development promote cooperation in research and development, a new theme in Japan's industrial policy. For instance, JETRO lauded Nokia's research center in China responsible for development of low-priced cell phones. Officials propose a fundamental shift in attitude at Japanese firms: "[T]he time has come to shed insistence on acting alone and to take advantage of technologies, know-how and human resources from around the world (JETRO 2006a: 31)." Given Japan's penchant for importing foreign technological advances for local product development, one might wonder what is different here. Market alliances between leading firms at home and abroad does not suggest a reorientation toward more comprehensive collaboration in a new "knowledge" nexus.

Open innovation networks represent one path for recovering prominence in regional production chains. But innovation would demand new patterns of cooperation with local firms, new channels of knowledge transfer, and more effective protections for intellectual property rights as well. The following chapter reviews innovation systems in Japan's three production partners, before turning to sectors of electronics, autos, and textiles.

4 National interest versus regional innovation

Japanese officials have high hopes for technology leadership in the regional economy. State planners in 2006 declared their ambitions for Japan as the "world's innovative center," contributing to the technological development of the East Asian region (METI 2006a). The goal is reminiscent of Akamatsu's "flying geese" thesis of Nikkei networks filtering technology down the various layers of production sites across the region in what Japanese officials considered to be a mutually beneficial pattern of industrial growth. Changing patterns of trade and investment have pioneered new types of knowledge exchange in the region. Some question how "benign" this new nexus of trade and investment might be for Japan and her partners.

Will national interest in East Asia trump cooperation in innovation? States remain committed to national development priorities, often organized around national innovation systems, and the ability to create and protect a nation's technology has long been linked to national security. Yet the wider project of innovation today can prosper only with multiple, diverse contributions across national borders. Market integration holds out promise for regional economic prosperity. Corporate stockholders, now major stakeholders in regional cooperation, may need to nudge reluctant state offices to recast national interest in more inclusive, market terms. Open innovation networks will demand a new consensus on technology and national interest, and new complementarities in local business systems to link production partners across national borders.

This chapter looks to national innovation systems in the four nations – Japan, China, South Korea and Thailand – to assess the new interplay of national interest and regional innovation. I argue that complementarities among firms, associations, and state bureaucracies in the region can promote cross-border knowledge flows through harmonizing standards and legal systems, and through new forms of sectoral cooperation between industries.

Division of labor

Critics of the benign Nikkei nexus countered that the new regionalization of production in East Asia by the early 1990s was based not on local needs and resources, but on the technology and network demands of Japanese firms (Morris-Suzuki 1994). It was less a benign nexus of mutually beneficial production than a "supply architecture" centered in Japanese innovation of critical components reinforcing a Japanese technological hierarchy (Bernard and Ravenhill 1995). Apart from questions about local benefit, the suggested leadership role had also faded. Even officials at Japan's METI admitted that the "flying geese" metaphor had lost its relevance in a region where industrial location no longer depended directly on a nation's level of economic development. METI officials acknowledged China's industrial clusters in labor-intensive industries such as textiles, and technology-intensive information industries broke the pattern and threatened Japan's regional hierarchy (2001: 17).

Technological dependence on Japan persists in some sectors nearly a decade into a new century, but with fundamental changes in the wider landscape of offshore investment. Global integration of markets across borders has progressed from production to innovation as MNEs increasingly invest abroad in local assets of human capital and research infrastructure (Castellani and Zanfei 2006). Technology cycles have been shortened and product development sequences accelerated. Innovation and technological change have been recast as short-term targets rather than long-term goals, critical for economic performance (OECD 2000). Three other changes deeply affect patterns and resources of innovation for Nikkei networks within the Asian region. One is the diversification of trade and technology links among leading Nikkei production partners such as South Korea and China, easing dependence on any single source of technology.

A second difference is the rise of China as a leading site for manufacture and now marketing, competing with Japan as the center for a regional division of labor. Some see a shift from a neat formation of "flying geese" to a chaotic flock of "swarming sparrows," buffeted by a strong Chinese "crosswind (Felker 2004; M[illegible]on 1991)." Apart from China–Japan competition, diverse development paths in different sectors is a third difference evident in the new division of labor. Production in electronics, for instance, has shifted from export of parts to offshore "screwdriver" factories, to location of production according to the degree of value-adding, with a technology ladder of contributions from multiple sites (JETRO 2007a).

T. J. Pempel pointed to regionalism or "regionalization" as a further feature in the rapidly evolving East Asian region (Pempel 2005: 19). Preferential trade agreements herald a new architecture of bilateral and regional divisions of labor within targeted sectors, and new complementarities of firm and market to support wider cooperation (Dent 2006). Market integration in the ASEAN countries has advanced the flow of parts and assembly facilities across member states. Geographical proximity and strong trade links among three nations in Northeast Asia with a remarkable concentration of global facilities for electronics manufacture has supported further integration. Examples of the latter include Denso's South Korean subsidiary, Denso PS, producing auto parts in Tianjin for Korean automakers in China. Toray Saehan in South Korea established a non-woven "spunbond" fabric subsidiary in China in 2006 (Toray 2006; 2008b). A choice between techno-nationalism or more open modes of cooperative learning abroad may well determine the fate of regional integration.

Whither Japan?

What is Japan's role in the welter of regional transformations? Ozawa argued that Japan had been the catalyst for structural upgrading as partners emulate and compete in the region, functioning as a "vital intermediary" of technology processing and transfer (1999). Given the prominence of Nikkei production networks in South Korea and now in China, Japan's "intermediary" role appears to have been at least one major factor in their technological progress, although the same role has not led to similar technological progress in Thailand. If there is at least some precedent of technological learning in Nikkei regional production networks, it strengthens the prospect of more extensive forms of collaboration with subsidiaries, affiliates, and independent sub-contractors, as well as with Asian competitors. But the possibility that learning can benefit both home and host firms does not resolve the dilemma of narrow national interest in a zero-sum regional economy, versus regional cooperation in a positive-sum market.

Precedent permits a more sober assessment of prospect, and the question here is not simply one of precedent but also of prospect. Can Japan recover her indirect, largely economic form of soft power in a regional division of labor increasingly focused on China? Can Japan develop the institutional complementarities among her production partners necessary to tap local resources of experience and know-how? And finally, can Japanese investors compete with "asset-seeking" Western investors in Asia, globalizing their own innovation networks?

Interest and innovation

Technology can serve as an instrument of national power in a global economic competition. The role of knowledge in national security has been widely debated in the literature on techno-nationalism (Simon 1997). Richard Samuels brought attention to Japan's ideology of technology protection as being fundamental to national interests, a strategy of indigenizing, diffusing, and nurturing technology to make the nation rich and strong (Samuels 1994). He concluded that techno-nationalism, as opposed to techno-protectionism, will be part of any "intelligent" nation's security strategy as a means to improve the skills and ensure the welfare of their citizens. The argument did not dampen continuing criticism at the time of Japan's reluctance to transfer technology in her own production networks (Hatch and Yamamura 1996; Mochizuki 1995). Nor does the conclusion answer today the question of effective borders of nationalism in knowledge flows no longer closely bounded by the geography of nation-states.

A new element has been added to the debate recently with Japanese officials publicly faulting the nation's firms for their reluctance to collaborate across borders (JETRO 2006a; METI 2006a). Officials extended the criticism the next year to the reluctance regarding the outsourcing of production, much less the sub-contracting of development (JETRO 2007a). Clearly this was not a call for regional cooperation in knowledge flows, but simply for technical cooperation with advanced economies on common global issues. Some critics dismiss the effort due to its self-interested focus on the advancement of Japanese firms. I would argue, however, that the growing number of non-Japanese technicians in state-funded research units, and increasing concern for global issues in state policy augurs well for state tolerance if not support for firm-level cooperation across borders (Schaede and Grimes 2003).

Japan's caution in opening the borders of her own innovation system may well be affected by the persisting techno-nationalism of her neighbors. Chinese state officials are not shy in the promotion of nationalist goals in technological development, an ideology widely espoused in South Korea's own development decades from the 1960s (Segal 2005). The fact that production partners such as China and South Korea share Japan's commitment to technology as a means of national security may discourage deeper collaboration in the long term, just as anxieties over intellectual property have already impeded Japan's collaboration with China.

Borders

Yet other trends favor cooperation. Permeable national borders of technology, and the gap between interests of governments and globalizing firms in East Asia have intensified over the past decade with expanding ties of trade and investment, and formal political efforts at economic integration. Keohane and Nye argued that a more decentralized, heterogeneous world had emerged in place of an international system based solely on the sovereignty of nation-states (2000). The growing complexity of multiple global networks of interdependence now plays a major role in tandem with national governance. Any effort to define "interests" in international trade and investment must include systemic or region-wide actors and organizations, state or national actors and interests, and corporations and their associations.

Regional trade arrangements at once reflect and reshape global value chains within Asia. For instance, business associations in East Asia play a major role in defining "national competitiveness" in negotiations of preferential trade arrangements. Growing interdependence among these levels of region, nation and enterprise has deeply affected the content and articulation of "national interest" in technology policy. Boundaries of national interest have not deterred major Japanese firms such as Panasonic and Denso from alliances with foreign firms in technology development (Brown & Co. 2001; ED Research 2007; Osono 2003). How the emerging nexus of national interest, innovation and regional integration will develop remains to be seen. I look here to national innovation systems, and in subsequent chapters to innovation and regional integration at firms such as Panasonic, Toyota and Denso, and Toray and Itochu.

National interest versus regional innovation is a question of principle or priority. Harmonization of standards and even coordination of innovation across borders is a question of practice, perhaps in time a question of convergence at the corporate level to global "best practice." The pressing issue of evolving knowledge flows in production networks is thus a question of both why and how. How can expanding trade and investment, as well as trade liberalization within the region, promote innovation across borders? Precedent in Japan's technological learning process and that of her production partners offers one window on prospect or at least potential for collaboration. Production networks in textiles, automobiles and electronics have spurred technology transfer within Japan's domestic economy, but can this process be replicated among her production partners? The goal is mutual learning, a regimen through which pertinent information becomes useable knowledge, often stimulated through collective learning within industrial clusters (Camagni 1991).

New spaces

A distinctive space of place and space of flows distinguishes technological learning in East Asia. For instance, the region provides a home for nearly three-quarters of installed capacity in man-made fibers. The Asian Federation of Chemical Fiber Manufacturers acts as a hub for information flows on supply and demand among national associations of chemical fiber mills, and technology advances. Place includes geographical location, often in clusters of similarly specialized firms or sectors. The value chain extends across multiple locations spanning national borders, where technological learning among firms, or affiliates within a firm, proceeds across different national systems of innovation. A space of flows includes the information exchange in manufacturing and product development with the GVCs and their sectoral industry associations. Much work on innovation looks only to the final stage of knowledge-intensive production, but earlier, preliminary phases of innovation can be found in labor-intensive and capital-intensive production as well. I look to technological learning stretching from more labor-intensive to more capital-intensive sectors. Indeed, Nikkei networks in East Asia remain mainly in the first two stages, with limited investments in South Korea in knowledge-intensive projects.

Technology includes not only human skills, but also institutions of firm and association, and wider social structures such as national innovation systems. It refers to abilities within firms and sectors to absorb, adapt, and advance "existing know-how and skills (Hobday 1995: 32)." What results is a variety of innovations such as effective operation of available technology, minor adjustments of machinery or process that improve the productivity of the technology, or major innovations in the design and purpose of the technology. Innovation is not simply what can be imagined and designed, but also what can be profitably manufactured and successfully marketed. We find an interactive, iterative emphasis recently superseding the linear image of invention, product development, and diffusion into the market. This "iterative process" approach to innovation highlights the need for effective interaction among different sections within the firm, or different firms within the cluster, or among different segments of the GVCs.

Much attention is paid to *kaizen*, a specific system of incremental innovation or continuous improvement involving managers and workers alike. Japanese-style innovation has been characterized as incremental, a complex melding of skills (*suri-awase*) within firms and closed networks of firms, significant for both results and the learning process (Herbig 1995). But just as Japan caught up with the Western economies, those economies were shifting to a new phase of "industry-based knowledge and open networks (Jones and Yokoyama 2006)." An OECD report found Japan among the least active nations in international cooperation in R&D.

Strategies for learning

If the goal is mutual learning, how does the process of offshore production promote local learning of foreign expertise, as well as permit investing firms access to local know-how and contextual knowledge? Two related modes of value chain participation encourage local learning. Sub-contracting for Nikkei firms permits access to international markets, standards, and production and management techniques. Original equipment manufacture (OEM) sub-contracting has played a major role in the development of East Asia as the world's factory, particularly in electronics and textiles. In contrast, foreign direct investment (FDI) brings a Japanese firm to the host country, "transferring a package of capital, skills, technology brand names and access to established international networks (Guyton 1996: 180)." Other less direct benefits can also be significant, such as spillovers to local skill creation, and a stimulus to competition among local firms.

Package versus process

Technological development in Asian NIEs such as South Korea, but also in China, make possible higher-value-added production in Nikkei production networks in the region. It also permits more efficient specialization of tasks in the development of products within a value chain, a more sophisticated division of labor. As rising levels of consumption in the region contribute to larger domestic demand, Nikkei networks can focus on domestic markets as well as exports. The reorientation requires far more extensive access to tacit forms of local knowledge such as consumer taste and effective marketing techniques. But clearly the role of local knowledge and capabilities differs dramatically not only by country and sector, but also by the local or export focus of the foreign firm, and their own capabilities of both teaching and learning.

Intriguing questions of technology "package" versus technology "process" offer additional markers to distinguish the Asian production networks. We noted above the link between open versus restricted investment regimes and access to foreign technology. Package and process move the analysis from FDI profiles to corporate practice within industrial sectors. A "package" approach refers to a technology transfer through equity participation. Foreign partners procure machinery and convey the know-how in operations, management and marketing to insure profitable use. Advantages of rapid transfer must be balanced with possible negative scenarios of technological dependency resulting from closely held "black box" technology transfer from the home country, and a passive approach of simple machinery operation rather than learning in the host country. The box analogy suggests critical components secured against unauthorized duplication. Possibilities, however, of "leapfrogging" the learning process in some sectors may be worth the risk. A successful transfer of state-of-the-art technological packages among collaborating home and host country firms can promote rapid technological development, well beyond the more deliberate pace of learning by trial and error.

Liberal investment regimes in China and Thailand welcomed inward foreign direct investment (IFDI), though with restrictions to insure local participation in joint-ventures. Foreign ownership and management permit more rapid development of sectors. The prospect of eventual participation in China's huge domestic market attracted growing numbers of foreign investors once procedures were clarified and restrictions on production and market access eased. With a much smaller market and largely agrarian economy, Thailand did not attract a similar variety of industries or investors beyond Japan and the USA until recently.

Chinese firms have followed the package transfer, usually through joint-ventures with foreign firms rather than "turn-key" transfers of plant and machinery. One advantage of a Toyota engine plant in Tianjin, for instance, is that joint-venture partners at First Auto Works get to observe the organization and management of the process. They also have the opportunity to "unbundle" the package over time in the operation of the plant, assuming again good personnel with the time and talent for learning. Japanese FDI to Thailand has likewise followed the package approach, but without the results of rising technological levels apparent among their northern neighbors. A review below of national innovation systems tells us of local capacity for capitalizing on knowledge flows, but little of Japanese efforts to convey that knowledge.

Learning in restrictive investment regimes taps foreign knowledge streams without permitting foreign ownership. Japan and South Korea long insisted on indirect foreign investment, using licensing rather than foreign participation in the nation's infant industries (Eichengreen 2002). OEM production was another example, with designs and specifications of the "process" coming from abroad, but with management, ownership, and technology all in local hands. Exceptions were permitted at initial stages of critical sectors in the formation of South Korea's "full-set" industrial structure, but even here local management and operations quickly replaced foreign personnel. The exceptions deserve closer scrutiny.

A South Korean precedent of technology upgrading from inward investment suggests that nations committed to indigenizing industries can cull *process* from *package*. More to the point, the Korean track record of learning "processes" suggests that a nation can make its own progression from technological process to industrial packages in a full-set industrial structure. Rather than technology as such, we find practices embedded in local institutions that promote learning critical in the industrial growth of both Japan and South Korea. The key is the ability to absorb imported process and product architecture to permit extensive technological learning of the imports, as well as subsequent adaptation and improvement of the process. But if the South Korean experience of technological learning offers hope for similar host nations in IFDI, Japan's own investment regime today does not augur well for learning in a home nation with low levels of IFDI.

Japan made initial steps to reduce restrictions on foreign investment already from the 1970s, but remains today well behind other advanced industrial economies in the stock and flows of IFDI. The stark contrast between Japan's failure to make significant local reforms in corporate practice and market structures to attract foreign investment and the remarkable progress China has achieved in structural reform to encourage foreign investment only promotes investment in China at Japan's expense. South Korea's liberalization of its own investment regime from the 1990s has already attracted considerable IFDI, again in contrast with Japan. Insulation of learning appears the critical factor in Japan's reluctance to enact effective reforms of its investment regime. Mireya Solis has written of the effort to "separate technology from ownership in order to acquire proprietary knowledge while at the same time limiting managerial participation by foreign MNCs (2005: 128)." "Defensive liberalization" both deflected pressure from global partners, but also accommodated a small number of major Japanese firms that recognized the benefits of collaboration.

Coordination

Differences of state coordination, industrial structure, corporate scale, and production methods deeply affect patterns of interaction in Asian production networks. States play a major role in innovation through formation and diffusion of standards and regulations, consolidating resources and establishing directions in national innovation systems, and direct research support. Japan, South Korea, and China all appear to fit a model of a "coordinated market economy," rather than the variation of a more liberal market model evident in Thailand since the Financial Crisis. The relatively tight coupling of business and state has been linked to vertical integration around major *keiretsu* in Japan, but appears evident in the industrial policy of South Korea and China as well.

Hall and Soskice wrote of "group based coordination" in Japan and South Korea, where *keiretsu* or *jaebeol* represent families of firms with dense interconnections (2001). Leading Chinese industrial conglomerates remain mainly state-owned corporations. Third, business associations play a prominent role in information exchange and sectoral standards in Japan, China, and South Korea, though less so in Thailand. China's industry associations have evolved out of state bureaucracies overseeing the sectors, and remain firmly in the hands of former state officials. Japan has strongly encouraged and in some sectors extended financial support for the development of Thai industry associations.

Industrial structure

Differences between a full-set industrial base versus a specialized industrial sector can be very significant in anchoring foreign invested enterprise. Japan, South Korea, and China have invested in full-set textile, automotive, and electronics industries with a wide variety of heavy and light industrial products. Thailand has evolved rather with "supporting industries" in a relatively narrow industrial base.

Textile sectors stretch from upstream, technology-intensive synthetics to spinning and weaving, and dyeing, and downstream to garments and fashions.

Automobile industries include major assemblers and their first-tier suppliers, stretching downstream into a second and then far larger third tier of suppliers in steel, plastics, rubber, electronics, textiles and other areas. Electronics and electrical machinery likewise require a broad base of industrial sectors contributing components and parts for final assembly. Electronics and garments however, include highly standardized sections or "modules" that can be assembled and tested without extensive knowledge of the process. Automobile manufacturing remains largely "integrated" with multiple components of different specifications, despite efforts to modularize production.

How significant is the local industrial base for foreign investors? Toyota's chief executive in Thailand recently lamented the lack of sufficient local Thai steel capacity, forcing imports from China and India (Sonoda 2007). In addition to materials, manufacturing partners are also a priority. Japanese MNEs often replicate a *keiretsu* form of vertical integration abroad, including not only subsidiaries and affiliates, but also a network of Japanese small and medium-sized suppliers to insure quality and delivery of parts and components (Okada 1999). "Localization" of procurement often means procuring suppliers from Japanese firms that have moved abroad with the Japanese MNEs, rather than from independent local firms. Countries with full-set industries have advantages of scale and scope in the competition between Nikkei and local, independent suppliers, especially once the technological gap has been narrowed.

Scale

Corporate scale also affects patterns of interaction. Despite the prominence of Japan's diversified conglomerates such as Itochu or Toray in textiles and fashions, Toyota in autos, and Panasonic in home electronics, SMEs play a major role in production and innovation. South Korea's family-owned and managed *jaebeol* and China's huge state-owned enterprises likewise represent some of the largest firms in Asia. Thailand would be the exception among its Northeast Asian competitors with relatively few conglomerates in manufacturing. Although SMEs represent the majority of enterprises and employment in all four countries, we do not find a Japan-style partnership of innovation or information exchange between larger and small firms.

With vertically integrated *jaebeol* internalizing so many of their functions, South Korean SMEs remained "low-technology, low-cost producers" through the early 1990s (Hobday 1995). China's SMEs had initially developed within state-owned enterprises without a role in design or innovation. A narrow industrial base in Thailand offers fewer opportunities for smaller-scale manufacturing. The vast majority of Thai SMEs remain in service industries. Among South Korea, China, and Thailand, only Korean SMEs appear to emulate the Japanese pattern of SMEs in offshore investment as well as exports. Parallels could be drawn with South Korea's automotive *jaebeol* or with China's state-owned enterprises.

Third, Schaede's earlier work on "cooperative capitalism" highlighted the role of trade associations in maintaining cohesion within sectors, and between state and private sector (Schaede 2000; 2003). Significant differences in the scope and autonomy of association activities in the three host nations affect their contribution to "cooperative learning" within the industry (Okada 1999). I look carefully to two functions of associations in the learning process. A primary function of local or regional associations in Japan is information exchange among SMEs, and promotion of technology learning, often between larger and smaller enterprise, and with state-supported research institutions or universities. Smaller enterprise can specialize, constantly improve manufacturing processes, and focus on applications, particularly in the initial commercialization of a product (INT 2007a). A second function is the role of major sectoral associations in promoting standards domestically or across the region. Among a range of technology support institutions, associations can play an important role in standards, sectoral policy, training, and information exchange on technological developments.

Collaboration

Large MNEs of home and host country can collaborate in strategic alliances, such as Japan's Sony with China's Haier Corporation, or Hitachi of Japan and Goldstar of South Korea (Ernst 2006a; Simon and Yongwook 1997; TN 2007e). Sanyo for instance recently announced sale of a major stake in its Thai refrigerator factory to China's Haier Group. Sanyo (Thailand) President Isutamu announced that Sanyo would focus on higher-value-added products, while transferring the plant to Haier in line with its five-year-old collaboration agreement. In the second case, Hitachi agreed to transfer production technology for 1 Megabyte DRAM (i.e., Dramatic Random Access Memory) chips to Goldstar in 1989, and then to import the products back to Japan. Both the Sony and the Hitachi cases reflect hierarchical, vertical relationships between Japanese technology leaders and production partners abroad.

A more common form of collaboration in Nikkei production networks in East Asia is sub-contracting, often with medium-size local firms. Some complain of Japan's preference for technology sharing solely within her own *keiretsu* structure projected abroad, or internalized across the region within the confines of the global production system (Guyton 1996). Although collaboration with local subsidiaries and affiliates may contribute to overall skill development within the local economy and stimulate competition, it can also leave the local independent supplier outside of the technology circle in solely labor-intensive production.

Benefits of collaboration abroad within the wider firm and affiliates include protection of property rights and high trust. Disadvantages can also be cited, such as a limited set of suppliers, and limited possibilities for raising technological levels of production without a major investment by the firm itself. One major disadvantage of procurement from local suppliers without development of information exchange and learning is isolation from local know-how and familiarity with labor and consumer markets. Yet apart from broader features of local state, industry, and corporate scale, there remains the critical issue of home country patterns of learning and their relative portability as firms move abroad.

System

The Japanese production system tends to develop generalists rather than specialists, a strategy demanding intensive knowledge-sharing on the shop floor as workers rotate among tasks. Refinement of designs in the transition from concept or discovery to production is another characteristic of Japanese innovation, as is the transfer of an idea or concept from one industrial sector to another (Herbig 1995). What drives the system is the intensive exchange of codified and especially tacit knowledge, usually on the shop floor. Japanese firms attempt in different ways to replicate the system abroad, particularly in worker-training programs. One author argued that the Japanese focus on know-how demanded more time and interaction with home firms than the emphasis of Western firms abroad on simply transferring operational skills through manuals (Yamashita 1995).

On-the-job training may result in more effective local skill formation, including repair and adjustment of machinery beyond simply operation of imported machinery. The alternative is reliance on foreign technicians dispatched for repairs as needed. Investment in local skill formation takes much longer and demands more extensive interaction with Japanese experts. Differences in production methods and personnel training may explain the comparatively slow transfer of technology in Japanese firms abroad, compared to Western firms. Yet the argument fails to address other differences in contract and control of information. Japanese firms have long grown accustomed to permanent employment and the resulting worker loyalty to the firms, although the pattern is slowly changing today in Japan. Japanese affiliates abroad, unaccustomed to high rates of worker mobility, remain reluctant to share critical know-how. Whatever the process or reasons, localizing the Japanese production system abroad remains fraught with problems, whether at firms, sectoral associations, or technology support institutions.

Complementarity

Collaboration depends in large part on structural compatibility among firms, sectors, industry associations, and national innovation systems. The *keiretsu* structure of vertical integration insures compatibility at home, and has the advantage of initial intra-firm complementarities within its networks abroad. Effective governance in turn depends on adjustment and extension of these complementarities among subsidiaries and affiliates abroad. A further task is developing reliable patterns of production with unaffiliated local sub-contractors abroad.

Aoki argued that institutions in one domain such as the polity or economy constitute a framework or "institutional environment" for institutions in other domains and vice versa. *Complementarity* then refers to the presence or efficiency of one group increasing the returns from the presence or efficiency of the other group. He reasoned that such synchronic interdependencies constitute a "coherent whole" where individual institutions cannot "easily be altered or designed in isolation (2001: 225)." For instance, large-scale enterprise in Japan, South Korea, China, and also Chinese–Thai conglomerates, reflects some institutional similarity, or "isomorphic" properties.

Complementarity is often sustained by place, particularly among smaller firms. One effect of geographical concentration or "agglomeration" and sectoral specialization is that it permits external economies and critical sharing of knowledge horizontally among networks of SMEs, or vertically in production chains dominated by the larger multinationals. Michael Storper highlighted "untraded interdependencies" among the most critical assets of these spatially based centers, i.e., "conventions, informal rules, and habits that coordinate economic actors under conditions of uncertainty (1997: 5)." The sharing of "tacit" or non-codifiable expertise demands trust or solidarity within the network beyond simply proximity, specialization and scale.

Turning to the context of exchange, Taylor and Leonard wrote of a "structural embeddedness" which incorporates firms into local, spatially anchored networks, and in turn facilitates information exchange and learning (2002: 2). The dynamic included the following features: reciprocity (recurrent transactions), interdependence (trust and mutual orientations), loose couplings (ability to shift partners), and asymmetric power relations (dominance and exploitation between unequal exchange partners). Reciprocity and some forms of interdependence and asymmetric power relations appear relevant for Asian SMEs in Nikkei networks. Loose couplings remain a problem for the investment of home firms in host firms abroad.

The prominence of smaller enterprise in Nikkei networks abroad turns attention to the role of scale in knowledge transfer. The organization, relative importance in manufacture, and internationalization of SMEs, however, reflect discrepancies between the role of smaller firms in innovation in Japan and South Korea, versus China and Thailand. "Flexible specialization" in labor and production among smaller firms in a post-Fordist era depends in part on links to the new space of local knowledge networks abroad (Piore and Sabel 1984; Sabel 1989). Yet such links are forged with difficulty if Asian SMEs remain largely offshore factories tied to home country networks without embedding in the markets and business community of the new locale. The dilemma for Nikkei networks is not production, but rather integration into local networks to capture advantages of learning to enhance flexibility in labor and production processes.

Business environment

A recurring policy theme of continuity in regional "business environments" is one strategy for building complementarity in support of Japanese firms abroad (JETRO 2004). Beyond Japan, the same term appears prominently in the Joint Study Group Report on trilateral investments among China, Japan, and South Korea (2004). Commerce can lead to community only with institutional continuities. Corporations chime in as well. Auto industry respondents to a JETRO Survey on investment sentiment identified state sectoral policy, industrial structure, and associational autonomy among major priorities affecting investment. The three issues serve as benchmarks for assessment of the business environments abroad in East Asia, a Japanese style of institutional arbitrage in the investment process. Nor do they overlook the infrastructure of targeted sectors in each host country or scale of market. Apart from contract and the rule of law, investment incentives and constraints, the tax system, and especially sectoral standards appear critical in investment decisions.

Regional organizations such as ASEAN, ASEAN + 3, and APEC permit Japan with multiple opportunities to campaign for common standards in production and trade. Toyota and Denso championed an early program for harmonization of import standards with ASEAN, the "Brand-to-Brand Complementation Scheme." A common standard for contract and law in the market remains the basic condition for Japan's plan for a "seamless business region" in East Asia. Protection of property rights and consistent enforcement remains a problem for Nikkei networks in China. Industry associations offer a forum for the formation and enforcement of standards in textiles, autos, and electronics. But beyond scale, standard and association, national innovation systems represent the best single indicator of adaptive capacity, along with the ability of a host country local firm to transfer its tacit knowledge into codified knowledge useful for the home country firm. Of particular interest here are the "technology supporting institutions," in each country, and their role in each of the three target industries.

China: packaged learning

Japanese firms in China have flocked to south China in numbers rivaling their massive penetration on the periphery of Bangkok. Close to 600 firms were located in Guangdong alone in 2004, often developing operations from a regional base in nearby Hong Kong. Flourishing auto and electronics sectors have made Shanghai the most popular manufacturing site with nearly 1,500 Japanese firms, and 240 more in Zhejiang Province, and 700 in Jiangsu Province. Shanghai, plus Zhejiang and Jiangsu provinces of the Yangtze River Delta together form a base for the majority of Japanese firms in China. Zhejiang Province is a major center for electronics, and Jiangsu Province has emerged as a major textile center, including vertically integrated production lines for Toray and Teijin of Japan. Panasonic and Toyota have been the leading Japanese investors in Tianjin outside of Beijing, where a further 278 Japanese firms are now located (JSBRI 2006: 85).

National innovation system

Among nearly 500 Nikkei SMEs in China, respondents to a METI Survey on Business Activities in 2004 evaluated the comparative strength of their technology at home and in China. A slight majority (57 percent) judged the levels as being about equal, with most of the rest evaluating their technology at home as superior (JSBRI 2006: 88). Larger firms with a variety of technology levels can better target machinery and research for the Chinese market. Panasonic and Toyota have opened their own research institutes to monitor the market, to track supply of engineers and researchers, and for their own product development. It is not simply the scale of the domestic market and availability of skilled labor, but the current research commitment and future innovation potential that draws foreign investors.

A nation with almost a billion researchers and a total research expenditure of $115 billion has few competitors, apart from the USA and its neighbors of Japan and South Korea. A healthy share of high-technology goods in national export totals reflects the contribution of foreign firms to the local economy, and an opportunity for learning from top firms across the world (Lemoine and Ūnal-Kesenci 2004) (Table 4.1). Education data indicate the major strides in literacy and secondary education for a developing nation of 1.3 billion people (Table 4.2). High enrollment in secondary education, and now nearly 20 percent of the college-age population in higher education bodes well for the absorptive capacity in technology transfer. Resident filings for patents are not only high for a developing nation (Table 4.3), but the recent trajectory of growth is remarkable. In the six years from the baseline of 2000, China's annual applications to the World Intellectual Property Organization (WIPO) have grown by 212 percent. China's Huawei Technology Corporation ranks in the top ten in number of applications from firms in developing countries (Fan 2006; WIPO 2006b).

What I find most remarkable is not simply the applications but the actual number of patents granted, with China capturing fourth place globally already in 2005 (Table 4.4). The fact that foreign firms are responsible for 61 percent of the patents granted reflects a persisting gap between technological levels of foreign and local firms, and stands in sharp contrast to the very small contribution of foreign firms to Japan's total number of patents. Foreign-invested firms were responsible for 58 percent of China's exports in 2005, one-third of total value in manufacturing, and employed some twenty-four million workers. But Japanese business circles are keenly aware of state efforts to transfer technology from these foreign firms, to upgrade production, to localize production, and even to discriminate against imports in favor of locally made products (Beijing Center 2006: 22). A strong state commitment to improving the level of local technology, coupled with public investments in education and research, provide strong local support for "cooperative learning" between foreign and local firms.

Table 4.1 Technology in four nations, 2005

Nation/ Technology	*1: Researchers*	*2: Research expenditure ($US billions)*	*3: Research expenditure as % of GDP*	*4: High technology exports as % of manufactured exports*	*5: Number of patent applications*
Japan	677,206	118.02	3.13	24	371,495
China	926,252	115.1	1.25	30	40,346
South Korea	179,812	31.6	2.89	33	76,860
Thailand	n.a	n.a.	n.a.	30	2,227

Sources for columns: 1–2: OECD 2006; 3: OECD 2007; 4–5: World Bank 2006b: Table 6.10.

n.a. = not available

Table 4.2 Literacy and education in three nations, 2005

Percent of population/nation	*China*	*South Korea*	*Thailand*
*Literacy rates**			
Female	87%	97%	91%
Male	95%	99%	95%
Secondary level enrollment			
Female	73%	93%	74%
Male	73%	93%	72%
Tertiary level enrollment			
Female	17%	69%	45%
Male	21%	110%	41%
Information technologies			
% of population with Internet access	7%	65%	10.9%
Broadband subscribers	1.6	24.7	0.2

Source: ADB (2006).

* Literacy rates averaged between 2000 and 2004. Secondary and tertiary school enrollments for South Korea and Thailand 2005, and for China 2004.

Table 4.3 Patent filings by residents in selected countries, 2004

Country/ Filings	*Resident filings*	*Resident patent filings per million population*	*Filings per $1 billion GDP*	*Filings per $1 million R&D expenditure (ranking)*
Japan	368,416	2,883.56	107.26	3.41 #2
China	65,786	50.75	9.37	0.71
South Korea	105,250	2188.96	116.19	4.40 #1
Thailand	670	10.52	1.41	0.48

Source: WIPO (2006a).

Knowledge flows

A recent study of Japanese overseas R&D activities distinguished demand-drive, research-oriented "innovative R&D," from local support-oriented "adaptive R&D." The former type of innovative research tends to be concentrated in nations with higher technological levels, and the latter adaptive research in Asian nations where firms concentrate on production. China proves an exception, however, with growth in both types of research. The percentage of Japanese overseas subsidiaries performing innovative R&D in China rose from 8.7 percent in 1996 to 13.2 percent in 2000, and adaptive R&D from 22 percent to almost 28 percent (Table 4.5). Such an increase can be partly explained by reorientation to domestic markets after an initial period of export production. The increase in innovative R&D is due in large part to pressure from the Chinese state to localize research, and the results of the nation's investment in education and technological development.

Collaboration demands trust, usually structured into reliable legal protections for intellectual property. China signed the WIPO Convention in June 1980. China joined the WTO only December of 2001, and signed the Trade-Related Aspects of Intellectual Property Rights agreement (TRIPS) (WIPO 2007). But Japanese firms large and small continue to complain of difficulties with law and contract in their Chinese operations. Close to two-thirds of Japanese firms in China cited a lack of transparency in the legal system in the annual JBIC Survey of Manufacturing Abroad (JBIC 2006a). A JETRO Survey of Business Sentiment in 2006 reflected continuing concern with China's underdeveloped legal system and legal procedures, intellectual property rights, and taxes. The recent emphasis on "indigenous innovation" in China's Science and Technology Development Plan has further complicated the landscape. But anxieties did not dampen directions of investment: over three-quarters of respondents reported plans to expand business in China (JETRO 2007fb).

Table 4.4 Number of patents granted in selected countries by residents and non-residents, 2005

Type/Country	*China*	*Japan*	*South Korea*	*Thailand*
Residents	20,705	111,088	53,419	62
Non-residents	32,600 (61%)	11,856 (9%)	20,095 (27%)	491 (89%)
Total (rank)	53,305 (100%)	122,944 (100%)	73,514 (100%)	553 (100%)
Rank	4	2	3	

Source: WIPO (2006b).

Table 4.5 Number of Japanese subsidiaries abroad performing R&D in selected countries, 2000

Host/Research	*A: Total no. of Japanese subsidiaries*	*B: Among (A), subsidiaries performing research*	*Among (B), subsidiaries doing innovative research*	*Among (B), subsidiaries doing adaptive research*
China	1,140 (100%)	300 (26.3%)	144 (12.2%)	89 (7.8%)
South Korea	529 (100%)	241 (45.6%)	126 (23.8%)	61 (11.5)
Thailand	834 (100%)	164 (19.7%)	59 (7.1%)	62 (7.4%)

Source: Shimizutani and Todo (2007).

Associations

One-quarter of smaller Japanese firms with operations in China report problems, but with only 5 percent of these firms taking legal action (Table 4.6). Firms report even basic problems of collecting payment, particularly among firms producing intermediate goods where fear of alienating critical customers often precludes direct action. Industry associations in other countries often provide a forum for discussing and possibly resolving problems in business transactions, but local industry and trade associations are often semi-official with close ties to the government, and foreign associations are closely controlled. SMEs in Japan have a rich array of local support in industry associations, cooperatives, and local chapters of the Japanese Chamber of Commerce and Industry.

But the Chinese government restricted national chambers in China to a single office until recently, although they tolerated informal trade "clubs" of larger Japanese firms (CJCCI 2005). Restrictions deprived most Japanese firms in China of local area representation. Sectoral associations such as the Japan Automobile Parts Industry Association (JAPIA) tend to be dominated by larger firms in Japan, and yet provide some voice with the Japanese government abroad to improve the business climate for larger and smaller members. The absence of a strong, independent counterpart in China relegates both JAMA (Japan Automobile Manufacturers' Association) and JAPIA to information-gathering activities.

Packages

An infrastructure of learning in China, coupled with extensive inward FDI in manufacturing, augurs well for the transition from packages to localized process. The rapid rise of resident patents alone would not indicate learning, but state funding for research, investment in education, and the proliferation of research institutes in foreign-invested enterprises suggest that the transition to mastering process is well progressed. Well-publicized investments such as the Haier Corporation's research institute in Japan, and Lenovo's purchase and successful operation of

Table 4.6 Nikkei firms' intellectual property right problems, and responses in selected countries, 2002

Country/Response	*Problem LEs*	*Problem SMEs*	*Action LEs*	*Action SMEs*
China	18.1%	25.5%	8.6%	5.2%
South Korea	16.3%	21.1%	9.2%	10.5%
Thailand	12.9%	6.9%	7.3%	4.2%

Source: JSBRI (2006: 113).

* LEs = large-scale enterprises; SMEs = firms with fewer than 300 employees and capital up to only ¥300 million.

** "Problem" = considerable problems, but ignored; Action = some problems, and action of some kind taken (lawsuit or administrative action, etc.)

IBM's personal computer unit suggests that China may well be moving toward transplanting of its own packages abroad.

Yet the data also highlight continuing difficulties with state coordination of the market that will impede the quality of information and technology flows critical for moving from simply importing packages to localizing the process. Foreign manufacturing firms have long experience in protecting their most precious technology, even as they produce their products abroad. Continued evolution of China's legal infrastructure to insure transparency and consistency will be one condition for more extensive knowledge flows.

Inward FDI flows from Japanese corporations through 2006 have brought Japanese products and production processes into the forefront of electronics and auto production, though with stiff competition from Western and now South Korean investors. Whether Japanese firms can take full advantage of local knowledge flows, and develop knowledge networks in China remains to be seen. Cohesive networks binding the innovation systems of the two East Asian centers would spur regional integration and national prosperity in the two nations if the evolution of comparative advantage could be managed smoothly.

South Korea: from package to process

Japanese manufacturing firms investing abroad annually ranked South Korea among the top ten most attractive destinations for investment between 2001 and 2005. They cite market scale and growth potential as the major reasons, but also cite problems such as intense competition in the local market, and rising cost of labor (JBIC 2006a). Indeed, the ranking does not amount to much if the top two or three destinations get the lion's share of the investment. The bulk of Japanese investment in China and Thailand goes to the two sectors of electronics and autos. Investment in South Korea has traditionally been more diverse, in part due to the rapid development of strong local firms in both sectors (Fung and Iizaka 2004). Japanese joint-ventures in South Korean textiles and auto parts, as well as limited investment in the electronics industry, offer a footprint for Japanese FDI in Korea quite different from that in China or Thailand. Demand of semi-conductor and LCD firms for advanced components from Japan has brought some new Japanese investment recently (JSBRI 2006).

National innovation system

South Korea's commitment to innovation is perhaps best symbolized in a recent name change at the state's leading office for industry. The Ministry of Commerce, Industry and Energy (MOCIE) was renamed the Ministry of the Knowledge Economy (MKE) in 2008. Critical elements of South Korea's rapid technological development include a state policy and national innovation system, the large family-owned and managed industrial conglomerates or *jaebeol*, and access to foreign technology and technicians. Technological evolution has been both rapid and extensive, penetrating deep into society in a relatively few decades.

Seth documented how national investments in education have supported high levels of enrollment even at the tertiary level (Table 4.2) (2002). Investments in broadband connection and Internet access across the breadth of a very well-educated population have ushered in a new era of rapid, extensive information exchange. Research expenditure as a ratio of total GDP reached nearly 3 percent, ranking second worldwide only to Japan (Table 4.1). High technology exports by indigenous firms such as Samsung and LG add up to nearly 33 percent of total annual export values, though the share of Japanese components remains prominent as evident in South Korean imports.

Other nations have made major investment in similar infrastructure. But do research expenditure and national commitment to trade and manufacture in high technology lead to innovation? South Korea ranks third worldwide in total patent filings, and leads the world in patent filings prorated according to GDP, and according to R&D expenditure (Table 4.3). As for patents actually granted, South Korea ranked third worldwide in 2005, with better than one-quarter of the patents granted to non-residents (Table 4.4). The non-resident share of patents in South Korea thus falls between China's high ratio of 61 percent, indicating heavy dependency on research by foreign-invested firms, and Japan's low ratio of only 9 percent, reflecting relative insulation from foreign research and investment. Finally, what is the research profile of Japanese subsidiaries in South Korea? Among 529 subsidiaries listed in 2000, 24 percent invested in innovative research, 12 percent in adaptive research (Table 4.5).

Metanational innovation

The prominence of foreign technology, foreign firms, and technicians and researchers from abroad in the development of South Korean autos, textiles, and electronics has been described as "meta-national (Asakawa 2007; Doz, *et al.* 2001)." The process in the electronics industry began with OEM sub-contracts in the 1960s and 1970s, which permitted extensive technical collaboration with the foreign client. Large Korean *jaebeol* then made the transition into adapting and transforming foreign technologies, assisted by strong technology support institutions, and a protected market for home appliances. By the next decade Samsung and LG initiated development of their own products by establishing research centers at home and abroad, drawing on the extensive local research base, and employing the "Seoul Express" of weekend moonlighting Japanese scientists (Cyhn 2001; Saneh and Goonatilake 1994). Other firms such as South Korea's Goldstar collaborated directly with Hitachi to gain a place in Korea's domestic semi-conductor hierarchy (Simon and Yongwook 1997).

In contrast to Japan's insulation, Taiwan and South Korea have electronics firms able to identify specialist knowledge regardless of nationality, and to draw on that expertise through sub-contracting, licensing, joint-ventures and technological alliances. Ironically, leading South Korean and Taiwanese firms initially drew on Japanese expertise, before turning to global supplier policies for equipment and materials. Some attribute relative industry autonomy today to state policy despite the prominent early role of foreign firms. Rhyu and Lee argued recently that state policies to promote localization of component production in the 1970s and 1980s, and then of end products from the 1990s in collaboration with Western firms, permitted South Korean electronics firms to break free of technological dependence on Japan (2006). While the authors recognize the continuing complementarity of the two industries, South Korea's reliance on critical Japanese-supplied key parts and high technology devices has prompted criticism of state policy (Mukoyama 2005).

Process

A successful transition from package to process in lower-value-added manufacture might have been more of an advantage for South Korea in the absence of China's rapid transition. Caught between the high technology of Japan and the very productive use of lower technology in China with its lower labor costs, South Korea must establish a niche between the two neighbors. Mastering process in higher technological production, whether in electronics, autos, or industrial textiles and fashion, will sorely test the national innovation system. Can South Korea draw on complementarities with leading firms in Japanese sectors to foster better knowledge flows?

Despite parallels in the structure of business associations, and even in technology support institutions, critical structural differences impede cooperative learning between Japanese and South Korean firms. Although the Korean state has given more support to SMEs in the past two decades, they remained largely dependent on the *jaebeol* through Korea's development decades, and even now remain without a role in either design-in strategies, or independent innovation. Business associations have played a major role in representing sectoral interests in government policy, but without the patterns of cooperation on information exchange evident in Japanese counterpart associations. As one observer concluded, "it is no exaggeration to say that there has been less intra-industry or inter-firm cooperation in Korea compared to the situation in Japan and Europe (Kim and Kim 2006: 38)." What some may find more disconcerting is the relative weak level of social trust evident in both inter-firm relations, and especially in labor relations (Lee Kong-Rae 2006).

As NIEs such as South Korea approach the technological frontier, they only welcome FDI that brings advanced technology, rather than the traditional FDI transferring older machinery from declining home country sectors (Eichengreen 2002). The rapid trajectory of technological learning in South Korean firms, particularly of Japanese technology in the formative years of its major export industries, might give potential collaborators pause. Cross-border technological collaboration among relatively equal partners may prove one viable avenue of joint-learning, evident in the cooperation today between POSCO and Nippon Steel (Kim Dohyung 2007). In the absence of relative equality, however, strong legal protections of intellectual property would appear to be critical. South Korea signed the WIPO Convention in March 1979 and has recently emerged among its most active members. South Korea joined the WTO from its inception in 1995 and was signatory to the TRIPS Agreement.

Few events have been more significant in South Korea's credibility among global investors than the OECD's invitation for Korea to join as only the second Asian member in 1996. A survey of intellectual property rights (IPR) among Japanese business reported a higher rate of complaints by both larger and smaller enterprises about problems in South Korea than in Thailand (Table 4.6). However, the higher rate of action taken on these issues in Korea than in either Thailand or China may indicate better avenues for arbitration and adjudication. The interpretation would be supported by the conspicuous lack of complaints about IPR infringement or the legal system in South Korea in the JBIC Survey of Manufacturing Abroad (JBIC 2006a). In addition to structural bases for trust in cooperative learning, there is also the benefit of collaboration with a very complementary partner. South Korea and Japan are the two most competitive economies in high technology in the region. Studies of the effects of a proposed FTA have highlighted significant complementary strengths (Inoguchi 2006; Urata and Abe 2005).

Thailand: package without local process

Thailand again offers a foil to the innovation systems of China and South Korea. The scale of the Chinese market and potential for innovation drive Japanese investment on the mainland, in the absence of complementary institutions. But basic problems with contract, confidentiality and IPR, and the absence of comparable industry associations continue to impede cooperative learning. South Korea's education levels and technological base appear the ideal complement for Japan's cooperative learning. Continuities in legal systems, state–business coordination, and association appear to support cooperation with Korea, but intense competition in similar sectors, the weak role of local SMEs in innovation, and difference in the social structuring of trust within firms versus between firms, impedes cooperative learning in Korea.

The impressive profile of Thai education and technology indicates a far different challenge for cooperative learning. Indeed the relatively strong share (30 percent) of high technology products in Thai exports, coupled with the absence of larger indigenous industrial firms in high technology industries, indicates the dominance of foreign-invested enterprise in the kingdom's manufacturing base (Table 4.1) (Kohpaiboon 2006). Thais well recognize the dangers of packages without process. The President of the Thai Export–Import Bank warned the nation recently that predominately foreign-owned high technology firms in the kingdom accounted for 64 percent of the annual Thai export value, and a large share of recent growth (Boontherawara 2007).

Technology support institutions

Enrollments in secondary education parallel ratios but not the numbers in the educational system of the nation on their northern border. China's enrollments represent a population base of 1.3 billion compared to only 65 million in Thailand. Tertiary education remains a major impediment for attracting foreign investment in Thailand. As reported bluntly in a recent electronics industry report: "Thailand suffers from a chronic skills shortage in engineers (Reed 2007: 202)." Apart from education infrastructure, technology infrastructure also lags behind regional competitors. Few nations can compete with the technological infrastructures of global leaders such as Japan and South Korea, and with technological investment and growth in China (Tables 4.3, 4.4). Yet Thailand stands together with South Korea and China as a major hub in Nikkei production networks, a major trade and production partner, though it must upgrade its supporting industries to anchor that investment.

Three groups play major roles in the local Thai economy: the public sector including government enterprise, the foreign-invested sector in export goods, and the local Sino-Thai family conglomerates. Yet a fourth group of Thai SMEs provide the critical support in manufacturing for foreign firms. These include local firms with up to 200 employees and a registered capital below 50 million baht (i.e., $1.4 million at 35Bt/$1), representing 89 percent of all local firms in 2001, 98 percent of the number of firms in manufacturing, and 76 percent of total industrial employment (Miyagawa and Manawapatand 2003; Regnier 2004). Few mid-size local firms can bridge the gap between the large local conglomerates, and the array of SMEs largely in commerce but also in manufacturing.

Apart from direct inter-firm ties, mid-size firms might also permit more solidarity in the business community. Peak business associations such as the Federation of Thai Industries or the Thai Chamber of Commerce have not been effective in either building consensus or exchange of information. It is the Thai Automotive Institute, for instance, that provides information and training for parts firms, as well as policy directions for the industry, rather than the Thai Auto Parts Manufacturers' Association. The Thai Textile Institute has focused more on standards and testing, leaving industry policy to the various sub-sectoral associations in the Thai Textile Federation.

One major issue here again is inter-firm trust and solidarity, a premise in the Japanese model of cooperative learning. Weak solidarity within associations leaves the state without effective partners for building consensus and implementing sectoral policies (Regnier 2004). Some fault the state for overlooking indigenous technological capability in the effort to attract foreign investment, first with import-substitution through the 1970s, and then export-oriented industrialization from the 1980s. Others cite the mismatch between technology support institutions devoted to R&D, testing and calibrating despite the more basic training needs of Thai manufacturing firms in machine operation, design and engineering (Intarakumnerd 2006; Saneh and Goonatilake 1994).

Process and the Economic Partnership Agreement

Radical differences in Thai state direction, industrial structure, associations and innovation systems with its North Asian competitors has not dulled the enthusiasm or profits of Japanese firms. A survey of Japanese firms operating over 1,600 affiliates in the ASEAN Four in 2006 reported strong sentiment for "strengthening or expanding operations" in Thai electronics and auto manufacture. Insofar as "localization" promotes transfer of managerial and operational technology, respondents cited problems only of staffing and insufficient technological capacity, rather than more basic issues of confidentiality prominent in China. Indeed, nearly 20 percent of respondents noted that localization of management was not even an issue, while 33 percent noted the same of procurement, and 26 percent found no issue even in localizing R&D (JBIC 2006a).

But Nikkei networks need to improve the lower-cost local supplier base rather than relying on transplanted Japanese suppliers. Japanese state offices such as METI and its trade office, JETRO, have long worked closely with the Nikkei business community to improve the level of training, standards, and even innovation. Negotiation of the Japan–Thai Economic Partnership Agreement (EPA) provides a remarkably detailed picture of how Japan plans to build complementarities abroad in support of its own investment.

The focus was industrial scale and sector, and the goal was transfer of know-how in both management and production. Japan pursues a wider strategy of regional economic integration through EPAs. The goal is partnership through comprehensive ties in development goals, rather than simply reduction of "behind the border" barriers to investment and trade (Dent 2006: 236). A JETRO survey already in 1989 highlighted divisions within sectors hampering effective networking, as well as the need for standards and testing (JAICA and Inoue 1989; McNamara 2002). Japan later provided critical financing and technicians to establish the Thai Textile Institute and the Thai Automotive Institute in 1998. Following the Asian Financial Crisis, the Miyazawa Plan (1998–2000) provided a roadmap and funding for SME recovery (TN 1999; Regnier 2005).

An EPA Task Force cited results of this initiative as evidence of effective cooperation: the Automotive Technology Building Program, Office of SME Promotion, SME Promotion Act, and the SME Management Consultant (SHINDAN) System (Miyagawa and Manawapatand 2003: 31). The proposed EPA would vastly extend the institutional cooperation in the sectors of textiles and autos, in an attempt to develop better sectoral institutions not just for training, but also for basic R&D. Cooperative learning structured through technology support institutions would improve the basic skills necessary for more efficient Nikkei networks procurement.

Conclusion

Our study to this point suggests progress in "disembedding" Japanese craft and learning from place and local practice (Giddens 1990). More problematic is the "re-embedding" of craft and learning in the interaction of host and home firm. Customs and practices identified with a particular village, region, or nation are disembedded with the influx of international trade and investment. Others have highlighted the extension of global patterns with the "penetration of world market rhythms to connect physically separate communities," and of "integration into the new time and space of world markets (McMichael 2004)." Contemporary society tears space from place as we spend more of our lives in interaction with distant or "absent" others, communicating from afar. As disembedded institutions come to join local practice with globalized social relations, local networks must adapt vertically and horizontally, globally and locally. Cooperative learning offers one answer to the question of how local practices can become re-embedded in global patterns (Pescosolido and Rubin 2000).

Cooperation

Japan's distinctive model of innovation prospered within a densely articulated society. "Cooperative learning" could be pursued within regional clusters, often in vertical organizations (*keiretsu*) tapping the strengths of both larger and smaller firms, and supported by a variety of technology support institutions. Finance is yet a further support, provided through Japan Development Bank and other local banks (DBJ 2006; INT 2007a). Rather than homogenizing market players, cooperation even within industry associations evolved together with competition. Learning abroad within Nikkei production networks has evolved in interaction with local systems of production and learning. Competition from rising regional powers, and with Western firms in the Asian region is pressing the shift from simply production to innovation networks. A new China-centered nexus of trade and investment continues to reshape the division of labor within Asian production networks.

A conviction that cooperation can coexist with competition provides a basis for community and market in Japan's techno-nationalism (Samuels 1994). Recent EPA negotiations with Thailand and South Korea suggest more. Compatibility can enhance constructive market competition. Compatibility begins with a supportive "business environment," and evolves into a "seamless web" for trade and investment in the region.

State innovation systems with their technology support institutions provide the necessary education and training, as well as supporting innovation. Industry associations play a key role at the sectoral level of information exchange, building consensus on directions of development, and shaping and implementing state sectoral policy. Within the structure of local industry, the technological levels of SMEs and their relationship with larger enterprise appears particularly important for effective "supporting industries" in production networks.

Linking innovation systems

Changes in the regional architecture of production networks both affect and reflect changes in local systems of innovation. FDI brings benefits of trade, employment, and tax revenues which can be directed to local education and research. A better-educated population and technology support institutions in turn attract mobile international capital. China's rise in the nexus of trade and investment parallels its investment and achievements in technological learning. Sustained economic growth in South Korea has supported remarkable achievements in education and innovation, and development of an extensive research infrastructure for research. Thailand as well has extended educational opportunities in tandem with economic growth, although innovation has been limited for the most part to foreign firms.

Learning within manufacturing networks depends on both knowledge and interests. Home country firms as well as host country subsidiaries, affiliates, and sub-contractors must have the capacity for sharing tacit knowledge, as well as the interest in exchange of information. Our review highlights the role of technology support institutions in Japan's strategies for effective knowledge flows in incremental innovation. We then examined their relevance abroad with a focus on complementarities in the learning systems of China, South Korea and Thailand.

Significant differences in levels of education and systems of innovation were evident among the partners, as were the technology levels and research activities of Nikkei firms in each nation. Among the most important findings is that levels of investment and trade do not correspond directly with institutional compatibility. Nikkei direct investment assets in South Korea remain behind China and Thailand, despite strong complementarity in intra-industry trade, structure and functions of industry associations, and legal systems.

Compatibility of legal systems and protection of intellectual property rights appears critical for sustaining a mutual commitment to cooperative learning. The goal is cooperation for more effective competition, not for a convergence depriving one or both parties of their comparative advantage. We found that complementarity of industrial structure also contributes to learning, particularly the innovation role of SMEs which appears so critical in the Japanese model. But even if interests can be insured through legal protections of contract, the new frontier of national versus regional interests, as well as nation versus corporate interests highlights the shifting terrain of national technology policy.

National interest can be redefined with open knowledge networks as a priority. Nikkei production networks can harness a regional nexus of trade and investment to national systems of manufacture and innovation. The question remains whether and how cooperative learning in Nikkei networks affects the national interest not only of Japan, but also of each of the three host countries. But neither nexus nor nation tells us much of actual knowledge flows. The profile of interests and actual innovation emerges rather in the actual practice of Japanese corporations abroad.

5 Electronics sector

Global modules and local minds

Consumer electronics have been an engine of growth for East Asian nations. Core research at home, and product development abroad appear the rule for most foreign firms manufacturing in South Korea, China, and Thailand. Improving knowledge networks in regional production chains would not necessarily take basic research abroad, but rather upgrade market and product research with deeper flows of knowledge across subsidiaries and contractors. I single out features of local sectors in each country that might promote or discourage knowledge exchange. I then look closely to Nikkei networks in each country to assess precedent and prospect for knowledge networks. South Korea, China, and Thailand have succeeded with remarkably different anchors in the value chains of Japan's electronics production. Can they now lock in innovation?

An export-oriented strategy of industrialization often begins with textiles and progresses to electronics, permitting initial entry into critical value chains across borders. A typical modular industry, production in the sector demands little of the host country other than a disciplined workforce and adequate logistics for importing components and exporting finished products. States provide generous incentives to anchor their node in the supply chains of major electronics manufacturers and marketers. Some have been remarkably successful. South Korean firms progressed from original equipment manufacture (OEM) for major foreign electronics firms to indigenous products and brands through a process of incremental learning. Despite South Korea's success at unbundling process from package (Cyhn 2002; Hobday 1995), the few "developing" nations that gain a foothold in the sector remain largely export platforms with some production channeled to domestic markets.

Innovation

Members of Japan's Electronics and Information Technology Industries Association (JEITA) keep at home production of "equipment requiring high reliability and quality", including semi-conductors, server storage devices, and electric measuring instrumentation, while expanding offshore production of popular products (2007). Member firms have made a huge investment abroad. Japan's offshore subsidiaries produced about one-third of the total worldwide value of electronic parts and devices in 2005. Production volumes in the Asian region would indicate at least a conventional transfer of technology, including training in the use of assembly machinery and the organization of production, development of a group of reliable local suppliers specialized in specific parts, and adaptation of the product for the domestic market. Panasonic ranges across the region with extensive market research and product adaptation at subsidiaries in all three target nations, complemented with efforts to establish regional standards beyond the firm.

A melding of local minds with the demands of assembly for global modules can differ markedly across the region. Major discontinuities can be found in the interplay of investment in manufacturing versus development, and the response of host nations to the challenge of anchoring both assembly and innovation. Panasonic networks in consumer electronics in South Korea, China, and Thailand, highlight various stages in the transition from assembly to knowledge networks. Japanese firms have excelled with integral production, but Panasonic has succeeded with modular production in basic household electronics, integral production in compact, high-value consumer electronics, automation, and now in building controls for lighting and other functions.

Innovation demands strong institutions. Innovation across borders demands institutional compatibility. Short-term investors, for instance, clamor for comparable legal institutions for protection of intellectual property rights. Long-term investors look for still more, such as institutional continuities in corporate structure and practice, sectoral association, innovation systems, and state policy and practice. Dent cited the latter priorities to explain Japan's insistence on partnerships rather than simply trade agreements (2006). Today's priorities have evolved out of historical trajectories. Trade liberalization coupled with historical patterns of Nikkei FDI in Asian electronics set the stage for cooperative learning in supporting industries (*susono sangyō*) nurtured across decades to develop complementarity with Japanese institutions.

Complementarity provides the societal context for relational networks within modular and integral value chains. Deeper social ties or "embeddedness" would not propel a major shift by itself from the dominant Japanese pattern of in-sourcing both production and development, but it would help to insure the trust necessary for more open cooperation in product development. Cooperative learning drove a productive alliance of larger and smaller firms in the evolution of Japan's electronics industry, particularly beneficial for major firms such as Panasonic. Can the same firm replicate the supporting industries model in the region? Partners might include subsidiaries, sub-contractors, and major strategic partners.

Knowledge networks appear all the more significant given the competition within the technology regimes of the fast-changing local electronics sector across the region. Nikkei firms face the challenge of Chinese firms possibly leapfrogging the technology ladder, and South Korean giants like Samsung and LG gaining the technological advantage. Rivalries old and new only add fuel to the competitive fire. Panasonic must compete with old foe Sony at home and the Sony–Samsung alliance in South Korea. Panasonic's competitors among Western firms in China, such as Nokia, have gained the admiration of even Japanese trade officials for their adaptation to the Chinese domestic market (JETRO 2006a). Brimble and colleagues documented recruiting at Seagate and also Microsoft of local engineers for open source research, formidable competitors for local talent and market share in Thailand (2007; 2006).

South Korea: complementarity without cooperation

Samsung's heated competition with Panasonic in flat-panel displays garners wide attention and admiration in South Korea. The nation has also excelled in semi-conductors with some 100 firms spearheaded by world leaders like Samsung Electronics and Hynix Semiconductor. Samsung and LG Electronics rank among the world leaders in shipments of mobile phones. Japan remains a very formidable competitor for South Korea's top electronics firms today, but played a partnering role in the origins of the industry. Panasonic (or "Matsushita," as it was then called) provided technical assistance in the early 1960s for transistor radio factories at Samsung, and joined with Anam Industrial in a joint-venture for manufacture of color televisions in 1973. Linsu Kim attributed South Korea's rapid sectoral development to both "strategic alliances" with foreign partners, and "reverse engineering" based on local "absorptive capacity (1997)." Others cited a progression from assembly of imported parts, to incremental local development of higher-technology parts, then through modified designs of existing products, and finally design and production of new products (Lee and Lim 2001). If foreign multinationals played a major role in the origins of the sector, the insistence of the South Korean state on local ownership of the industry, plus incremental learning during the next decade of OEM sub-contracting, fostered strong local firms by the 1980s.

Major *jaebeol* firms pulled up the industry with a process-intensive, high-volume strategy. They wisely invested revenues to gain foreign technology, sometimes even foreign researchers, including moonlighting Japanese engineers to develop in-house research capacity (Asakawa 2007; Kim 1997). Hobday accurately singled out South Korea among Japan's regional competitors already in 1995, recognizing a very similar path of government support, concentration in larger firms, and localization (Asakawa 2007; Hobday 1995). Higher value-added trade has become a South Korean trademark with high-technology exports of $75 billion in 2004 amounting to 33 percent of total export values, significantly higher than the comparable figure of 24 percent in Japan (World Bank 2006b). Electronics provides a major share of these exports. The question today is no longer simply competition in East Asia, but prospects for cooperation in knowledge networks.

South Korea ranked fifth among global exporters of electronic merchandise in 2006 with $103 billion directed abroad, amounting to 31.6 percent of total exports (Table 5.1). Imports of $60.5 billion represented nearly one-fifth (19.5 percent) of total annual imports. Trade begins with manufacture. South Korea's 7.1 percent share of the value of global electronics production ranks fourth in the world behind only the USA, China, and Japan. The nation's total annual value of electronics production totals jumped from $94 billion in 2004 to $112 billion in 2007. Components accounted for 46 percent of the total, and production in radio communications and

Table 5.1 Electric and electronics sector trade as percentage of world trade for target countries, 2006 ($US millions)

Trade/Country	*China*	*South Korea*	*Thailand*
*Value of electronics imports**	*255,364.9*	*60,530.3*	*31,592.5*
Total imports	791,460.5	309,379.4	128,584.4
Electronics/Total	32.6%	19.5%	24.5%
*Value of electronics exports**	*355,733.7*	*103,151.2*	*37,875.8*
Total exports	968,935.6	325,457.2	130,580.0
Electronics/Total	36.7%	31.6%	29%

Source: UN (2008).

* Includes the following HS 2002 SITC Codes: Code 8471, automatic data-processing machines including computers; Code 8469, electric typewriters; Code 8,470, calculating machines; Code 8471, automatic data-processing; Code 8472, other office machines; and Code 8473, parts and accessories thereof, plus Category 85, electrical machinery and equipment and parts thereof.

radar represented a further 28 percent (Reed 2007). Production and trade has brought the nation into the center of global value chains (GVCs) in the sector. Analysis of the sector reveals structural similarities with the Japanese sector, and potential for alliances in specific product areas.

Sector

Complementarity between South Korean and Japanese electronics begins with the prominent role of SMEs in manufacture. Despite the fame of Korea's big four – LG Electronics, Samsung Electronics, Daewoo Electronics, and Hynix Semiconductor – some 100 medium-size (i.e., 300 employees or less) specialized firms provide the backbone of manufacturing, together with a larger number of smaller firms. Concentration of capital, technology, and intellectual resources within the major firms has left the SMEs with little role in development other than basic manufacture (Kim and Kim 2006). Yamazawa noted that prior to South Korea's entry into the OECD, it was Japan that funded training programs for SME technicians to promote technology transfer and address Korea's trade deficit (2000). Training reinforces complementary structures and worker attitudes. Recognizing the contribution of the SMEs, the South Korean government has instituted special support programs to develop up to 100 SMEs into global competitors by 2010 (Reed 2007). Improved SMEs would then be better positioned to participate in product development with local *jaebeol*, and to cooperate with Japanese SMEs in technology transfer.

An asymmetric interdependence on Japanese suppliers of high-value-added components for more sophisticated electronic merchandise is a second feature. Japan remains the nation's major import partner, providing 18.6 percent of the total value of imports in the sector, though absorbing only 6 percent of exports (Table 5.1). Japan plays a major role in components, providing 9.6 percent of the value of computer parts, and 20 percent of electrical machinery and parts (Table 5.2). A bilateral deficit with Japan in the parts trade is balanced with a surplus in the global trade. South Korea recorded a robust surplus in the value of world trade in parts, with $8.7 billion of exports of computer parts versus $3.2 billion in imports. South Korea also managed a surplus in the trade of electrical machinery and parts, with exports of $85.5 billion versus imports of $52.6 billion.

Nikkei electronics firms on the peninsula employ a workforce of 10,000, the largest Japanese commitment in any manufacturing sector. Larger firms such as Panasonic and Sony operate manufacturing subsidiaries on the peninsula, but nothing comparable to their presence in China or Southeast Asia. The limited commitment of Japanese majors has not deprived South Korean electronics firms of a place in GVCs. On the contrary, general trading companies in Korea's larger *jaebeol*, as well as major shipping and transportation firms have established robust national links in GVCs, similar to the early role of Japan's trading and shipping firms in international trade. Multiple parallels can be seen in the electronics sector of the two nations. Indeed bilateral complementarities in export structure, industrial strategy and wider patterns of domestic industry bode well for knowledge exchange. Both nations have prospered with export-oriented industrialization, including textiles earlier and electronics now. Both nations have flourished with interdependent, full-set industries, including petrochemicals, steel, and machinery industries supporting production in electronics.

Table 5.2 South Korea's world trade in selected electronics*, and Japan share, 2006 ($US millions)

Trade source/ destination	*Value of data-processing machines (Code 8471)*	*Value of parts for data-processing machines (Code 8473)*	*Value of electrical machinery and parts (Code 85)*	*Value of electronic merchandise (Codes 8469–8473, 85)***
Imports from Japan	*297.1*	*311.8*	*10,589.9*	*11,305.8*
Total imports	4,481.3	3,246.2	52,622.4	60,530.3
Japan share	6.6%	9.6%	20%	18.6%
Exports to Japan	*589.6*	*238.2*	*5,730.8*	*6,610.9*
Total exports	8,511.3	8,817.9	85,576.8	103,151.2
Japan share	6.9%	2.7%	6.6%	6%

Source: UN (2008).

* HS 2002 SITC Codes: Code 8471, automatic data-processing machines including computers; Code 8473, parts and accessories for use in computers and other office machinery; and Code 85, electrical machinery and equipment and parts thereof.

** Includes the following HS 2002 SITC Codes: Code 8471, automatic data-processing machines including computers; Code 8469, electric typewriters; Code 8470, calculating machines; Code 8471, automatic data-processing; Code 8472, other office machines; and Code 8473, parts and accessories thereof, plus Category 85, electrical machinery and equipment and parts thereof.

A third feature of the South Korean sector is the prominence of strategic alliances with foreign firms. Production of storage devices is one area of cooperation. Hitachi–LG Data Storage was established in 2000, with Japanese partners holding a 51 percent share. Trade promotion officials at KOTRA explained, "the Japanese parent company would bring technical know-how to the venture, as well as international brand power supported by a global sales and service network," with the South Korean partner responsible for manufacturing (2008a). Another example is Toshiba–Samsung Storage Technology (TSST), which joined resources of optical disk drive departments in the two firms in 2004 into a curious authority structure. TSST is at pains to distinguish between two "headquarters," one in Kawasaki (Japan), and the other in Suweon (South Korea). The two headquarters operate with two separate administrations (i.e., "directorates"), and decide marketing issues through discussion with "relevant organizations for mutual consent (TSST 2008)." Whatever the difficulties, the alliance presses on. Samsung concluded a cross-licensing agreement with Toshiba of Japan in 2007 on interface specifications for their newest NAND flash memory chips (KT 2008a).

Sony joined with Samsung Electronics in a 50:50 joint-venture for manufacture in South Korea of seventh-generation liquid crystal display (LCD) flat panels already in 2004. Reuters reported a $1.8 billion expansion of the venture to produce eighth-generation panels (2008). Sony, Toshiba, and Hitachi have established manufacturing facilities in alliances with major Korean firms, but curiously, not Panasonic. Indeed, Asakawa found a broader pattern evident in Panasonic's alliance with Toshiba from 2001 for LCDs versus Sony's cooperation with Samsung. Toray's technology and investment in Panasonic PDP (Plasma Display Panel) Corporation is a further example (Panasonic 2007). He termed the Panasonic strategy domestic-collaborative, versus a Sony strategy of international-collaborative (Asakawa 2007). Panasonic discussions with Samsung on joint-market promotion of the next generation of computer storage devices, the Blu-ray players, even if successful would be limited to sales (KT 2008b).

South Korean firms find partners among major firms in Europe and the USA. LG Electronics established a joint-venture with Phillips of the Netherlands already in 1999 for production of flat-panel displays. LG recently announced a cross-licensing agreement with GE Consumer and Industrial to strengthen LG's position in the North American home appliance market (KT 2008a), and the list continues. Strategic alliances, whether joint-ventures, licensing or marketing agreements permit new knowledge flows in specific product areas, which may also have spillover effects through exchange of personnel, or contact with each others' suppliers and clients.

Apart from strategic alliances and their potential for more advanced research, the profile of trade and investment in electronics does not indicate a strong basis of subsidiaries or an extensive set of sub-contracting firms for research activities. Despite extensive trade in electronics between the two nations, South Korea is neither a manufacturing nor research node in the GVCs of major Japanese electronics firms. Japan does provide some vital components for South Korea's electronics GVCs, but is not the sole source of supply. The Japan Electrical Manufacturers'

Association (JEMA) reported forty-three South Korean subsidiaries of its member firms in 2005, though many are sales outlets.

With 60 percent of Nikkei offshore subsidiaries located in the Asian region, the Association has a priority of "reinforcing cooperation with other Asian nations (JEMA 2007)." Their goal is to expand markets through both production of goods and creation of "values," specifically efforts at international standardization in concert with the ISO. Within the home electronics sub-sector, JEMA promotes a strategy of domestic production of new and higher-value-added products, and offshore production of popular general purpose products.

Panasonic

Panasonic is indeed a global firm, with offshore subsidiaries generating 48 percent of annual sales of $76 billion in 2007, and employing over half of the firm's 334,000 employees (MEI 2008). The firm invested $4.8 billion in R&D in the same year. But of special interest here is the commitment to Asia. Panasonic has amassed assets of $10.7 billion in Asia, the highest of any region outside of Japan. The electronics giant operates seventy-five companies in nine countries across the region, including China, Thailand, and South Korea. The regional subsidiaries operate with a workforce of 82,000 and consolidated net sales of $77.19 billion in 2007 (Panasonic 2008a). Panasonic Korea remains largely a supplier of components and end products, unlike Panasonic subsidiaries China or Thailand.

The firm opened business in South Korea in 2000 with two wholly owned subsidiaries for sales of car electronics, living products, and system controls. Panasonic Electric Works (PEW) followed in 2003 with their own sales for production systems, information systems, lighting, home appliances, and electrical components. PEW acquired Sin Dong-A Electric in 2006 for manufacture of automation controls (TKS 2007b: 896, 908). Despite global commitments, and an extensive manufacturing presence in Asia, Panasonic appears more of a competitor than a major manufacturing and development partner on the peninsula.

Prospect for knowledge networks

It would be difficult for Japanese firms to compete on the ground in South Korea, whether in product development or market share, but Japanese firms might find the opportunities for joint-research in advanced technologies very attractive. Low levels of Nikkei investment may discourage this knowledge exchange. Japanese firms usually invest in local research projects through a manufacturing subsidiary, and usually close to the manufacturing site. Such a pattern would not be feasible in South Korea due to the small number of major Nikkei manufacturing facilities on the peninsula.

President Hatanaka of PEW forerunner Matsushita Electric Works distinguished between product development (*katihatsu*) and advanced research, the former developing products tailored to local needs, and the latter linking with local universities, corporations and research institutes in advanced technologies (MEW 2006a: 36).

But the innovation system in South Korea does not permit easy access of foreign firms to local resources. Government research institutes clustered in the Daeduk Science Town outside of Seoul supported transfer of foreign technology in the early development of heavy industries through the mid-1980s, rather than basic research. Hershberg and colleagues found that universities have long provided a steady stream of excellent scientists and engineers, but not a stream of industry-related research discoveries (2007; Sohn and Kenney 2007).

It has been firms rather than government institutes or universities that have taken the lead in industry research, retaining the major research resources within individual *jaebeol*. Aware of the problem, the South Korean government has been developing programs to better link corporate and national research resources. KOTRA reported a total of 11,800 firm-affiliated research institutes registered in 2005, the majority (52 percent) in the electric and electronics industry, with a much smaller portion (17 percent) in machinery fields. The report specified ways in which such centers at foreign firms could participate in national R&D programs, such as techno parks and other research-centered developments (KOTRA 2008c).

Changes in investment patterns and sectoral development might permit expansion on other levels of exchange whether through affiliations or cooperatives (*kyōryokukai*) of SME suppliers. Proponents of the Japan–Korea FTA cite cooperation in the electronics sector as one benefit of preferential trade ties. For instance, standardization of parts in a bi-national industry alliance would permit extensive cooperation among parts and component suppliers. Production of more sophisticated machine tools and manufacturing equipment for semi-conductors and other parts on the peninsula has thus far eluded South Korean efforts at localization and technology transfer (Kim Dohyung 2007: 13). South Korean firms might also benefit from new flows of IFDI. Japanese investment in manufacturing on the peninsula would both bring a new supply chain of components to the peninsula, and raise technology levels. Closer cooperation might encourage state and industry to upgrade South Korean parts suppliers, in part to anchor supply lines within Japan to stem the China rush among Japanese SMEs. FTA advocates press their argument against trade barriers as well. Market protection would only deprive domestic assemblers of quality parts, although South Korean firms may be reluctant to surrender what Lee and Lim term "latecomer advantages" of quality differentiation and lower cost production (2001).

China: cooperation without complementarity

If South Korea has gained renown among the "newly industrialized countries," China has moved to the front of the pack among "developing nations." But can China break the technological dependence on foreign-invested firms to move up the ladder of development in electronics? Mining and manufacture accounted for 41 percent of GDP in 2005, reflecting dependence on sectors like electronics, autos, and textiles for employment and productive investment (KNSO 2006). Electronics alone provided jobs for ten million workers, 13.6 percent of the total manufacturing workforce of 73.5 million. Generating about 12 percent of profits in manufacturing, the sector registered a similar proportion (12.5 percent) of the nation's total manufacturing assets (NBSC 2007a). Only the USA ranked ahead of China's manufacturing total of $177.1 billion in electronic merchandise in 2004 (KNSO 2008).

Electronics alone generate better than one-third of China's global trade. Sector export totals of $355.7 billion in 2006 amounted to 36.7 percent of China's global exports for the year (Table 5.1). Imports of electronic merchandise amounted to $258.3 billion, 32.6 percent of China's total global imports (Table 5.2). Within the multilateral global trade, electronics represents a major portion of China's bilateral trade with Japan. Imports from Japan amounted to 15 percent of total imports of electronic merchandise for the year, though sector exports to Japan of $25.2 billion represented only 7 percent of China's total electronics exports. As in South Korea's reliance on Japanese parts (Table 5.2), we find extensive Chinese import of components from Japan, with 14 percent of parts for computers, and 16 percent of electrical machinery and parts sourced from Japan (Table 5.3). One major structural difference in the import flows, however, is most likely an inter-firm versus intra-firm flow. South Korean firms obtain higher value-added parts from Japanese firms, and Nikkei affiliates in China import parts from mother firms in Japan.

Table 5.3 China's world trade in selected electronics*, and Japan share, 2006 ($US millions)

Trade source/ destination	*Value of data-processing machines (Code 8471)*	*Value of parts for data-processing machines (Code 8473)*	*Value of electrical machinery and parts (Code 85)*	*Value of electronic merchandise (Codes 8469–8473, 85)***
Imports from Japan	*802.8*	*2,706.4*	*35,226.3*	*38,815.2*
Total imports	19,924.5	19,142.4	218,965.7	258,364.9
Japan share	4%	14%	16%	15%
Exports to Japan	*7,383.3*	*1,717.2*	*16,010.4*	*25,296.6*
Total exports	93,017.3	33,033.2	227,476.3	355,733.7
Japan share	7.9%	5%	7%	7%

Source: UN (2008).

* HS 2002 SITC Codes: Code 8471, automatic data-processing machines including computers; Code 8473, parts and accessories for use in computers and other office machinery; and Code 85, electrical machinery and equipment and parts thereof.

** Includes the following HS 2002 SITC Codes: Code 8471, automatic data-processing machines including computers; Code 8469, electric typewriters; Code 8470, calculating machines; Code 8471, automatic data-processing; Code 8472, other office machines; and Code 8473, parts and accessories thereof, plus Category 85, electrical machinery and equipment and parts thereof.

Sector

Electronics has been central to China's export-oriented development strategy. One feature of the sector is the scale and structure of production and consumption. A full-set industry including steel and petrochemicals supports electronics, while complementary sectors such as construction and the automotive industry create demand. A booming sub-sector of consumer electronics has driven growth for the sector, with domestic demand now complementing demand from abroad. China has added to its status as the world's factory for "modular production" with the distinction of being one of the world's fastest growing markets for electronic merchandise. An expanding number of research centers among foreign firms, and rapid development of the national innovation system evident in patent activity noted in Chapter Four, suggest a further distinction as one of the world's leading research centers.

A second feature is the remarkable trajectory of technological development in the sector, and the prospect for further advances. Global players among local firms such as Haier in consumer electronics, Lenovo in computers, or Huawei in telecommunications have received wide attention. Huawei stands apart in scale, history, and innovation among the telecom equipment leaders. Established in 1988 by a former military officer, the firm reported profits of $512 million in 2006, revenues of $8.5 billion, and contract sales of $11.6 billion.[1] Among its 26,000 patent applications, the firm has gained approval of 4,200, including 7 percent of the Universal Mobile Telecommunications System (UMTS) patents for third-generation cell phone technology. In a recent study, Fan attributed the firm's success to corporate leadership with strong state ties and a research focus, an early start, access to foreign technology, and strong absorptive capacity (2006). The same author concluded that Huawei, ZTE (Shenzhen Zhongxin Technology), Datang Telecom and some others had defied the conventional learning curve by jumping ahead of the competition in telecommunications with their own technologies.

South Korea's rapid advance in telecom and semi-conductor sub-sectors surprised those who assumed that only incremental growth was possible. Scholars were forced to recognize the possibility of multiple paths up the technology ladder. Lee and Lim compared sectoral growth in automobiles and semi-conductors and concluded that those sectors had circumvented limitations of a unilinear, incremental process of catching up with advanced economies in the technology race (2001). Now China has likewise challenged assumptions about incremental growth, at least in the modular production of electronics, and technological advances in telecommunications. The scale and speed of sectoral development evident in extensive and reliable capacity for modular assembly, in an array of competitive local consumer electronics firms, and now in major telecom industry firms has led to a reassessment of innovation paths within GVCs.

Could market scale and access to external knowledge networks through anchoring of chains in countries such as China and India foster rapid rather than incremental technological advances in targeted sectors? Lest the exception be seen as the rule, Altenburg and colleagues singled out unique factors of the local context such as the role of the state as well as of foreign-invested enterprise, and the scale of China's domestic market (2008). China has not emulated the complementarities found in Japanese and South Korean models of sectoral development. Nonetheless, leaping up the technological curve in one sector raises the possibility of similar advances in other areas, greatly enhancing the significance of research clusters such as Zhongguancun in Beijing, or areas near the leading universities and research centers in Seoul, and adding urgency to the challenge of tapping new knowledge networks.

A prominent, indeed dominant role of foreign-invested enterprise is a third feature of China's electronics sector, driving much of the export and reaping much of the profit. In contrast to the development trajectory of Japan and South Korea, China welcomed foreign equity investment in the sector, initially for technology transfer and export programs, and later to serve growing domestic demand. Government policy for attracting FDI was one factor, but the relatively skilled and low-cost workforce, potential domestic demand, and proximity to existing offshore production bases in Asia added to the attraction. Foreign-invested firms in the electronics sector generated 61 percent of the annual value-added profit in 2006, and employed a similar proportion of the ten million workers in the sector (NBSC 2007a; 2007b). A recent joint-report by the OECD and China's Ministry of Science and Technology raised concerns over the dominance of foreign firms in the high-technology trade, particularly in information and communications technology, coupled with their reluctance to move beyond product development research on the ground in China. But as the authors point out, the national innovation system invests mainly in similar market and product studies, with only 6 percent of GDP of R&D devoted to basic or applied research, as opposed to experimental product development (OECD/MOST-C 2007: 15, 23).

How are the foreign firms tapping China's local knowledge networks? Certainly there is growth in the number of research institutes established by foreign-invested firms. Interest in localizing development facilities among foreign firms suggests a transition highlighted by Narula and Dunning from simply market-seeking FDI to investment in strategically related "created assets" supporting penetration of domestic markets (METI 2007b; Narula and Dunning 2000). Todo and Shimizutani found that the latter effort at tapping local research resources includes both "adaptive" R&D focused on adjusting process and product to local context, and "innovative" R&D focused on acquiring advanced knowledge resources in the country (2007).

Adaptive R&D was evident from a study of German firms by Chen and Reger in the Shanghai area with firms sharing a priority on localization of production and management systems. Firms worked to transfer management and quality control skills, as well as machinery operation and problem-solving, but not to transfer ideas through product prototypes or patent activity (2006). The results were echoed in interviews at China's telecoms contrasting higher-level R&D staff at local firms with marketing and product customization staff at foreign firms (Fan 2006). Other studies single out regional differences affecting R&D efforts. For instance, Motohashi and colleagues have cited "market-driven" research among Japanese firms in Shanghai and the Guangdong area, but more "technology-driven" research in Beijing with its cluster of scientific institutions (2006; 2005).

Panasonic

Few firms can compete with Panasonic's commitment to China, as noted in the *Economist* already in 1997: "No company has bet more heavily on the Chinese market and none has done more to prepare the ground than Panasonic, the world's biggest consumer electronics firm (1997)." Technology transfer permitted initial entrée for the firm. A former Panasonic China executive, Asada, cited some 150 plant and technical assistance projects led by his firm in the six years after Japan's Peace Treaty with China in 1979 (2006). The Shanghai Radio and Television Group, for instance, produced black and white TV screens with Panasonic assistance, and now is a partner in Panasonic PDP Shanghai (SFIC 2005). Ohashi added that Panasonic finally was able to get permission for a 50:50 joint-venture with a state-owned firm only in 1987 for color television manufacture, a sector dominated by Japanese firms through the 1990s (2006).

Panasonic invested heavily between 1992 and 1997 establishing thirty-seven separate manufacturing ventures, which have grown today into large-scale enterprises (JEMA 2005). Such a vast network of subsidiaries provides multiple opportunities for knowledge exchange, with major differences according to product. Deffree tracked a trajectory of product technology from household appliances in the 1990s to semi-conductors and printed circuit boards at their Suzhou operation, and now to the lighting and building controls in the construction industry (2008; MEW 2006b). The Panasonic Group listed seventy-eight subsidiaries in China by 2006, under a holding company in Beijing, titled Panasonic Corporation of China. The latter listed a paid-in capital of $478 million, and a staff of 1,600 employees (JAPIT 2007: 421).

Panasonic R&D in China addresses three critical areas for the firm: the need for input of local consumers to customize products for such an extensive and fast-changing local market; the need to keep ahead of the products of local competitors; and the need to tap local research resources better than Panasonic's Japanese and Western competitors. The mother firm operates five research facilities in China with a staff of 450 employees and a total paid-in capital of $25 million (JAPIT 2006; TKS 2007a). Established in 2001, the Beijing Laboratory today in the Zhongguancun area stands atop the firm's local research pyramid. Departments include mobile technologies and speech recognition, audio-visual development, and electronic components and functions (Panasonic 2008b). A center in Dalian specializes in software research, another in Tianjin near the Toyota plants works on automotive systems. A research center in Shanghai does studies of lifestyles and another in Suzhou provides support for the semi-conductor production plant (CD 2005). Panasonic joined with Japan's NEC and China's Huahei in 2001 to found COSMOBIC Technology Company for development of mobile phones and associated communications technology (COSMOBIC 2007). Larger manufacturing facilities also engage in product and process development to fit local conditions, in addition to the institutes.

Prospect for knowledge networks

Panasonic's knowledge network in China stretches across a vast chain of manufacturing in the electronics sector. Levels of research differ with product, but as the firm moves up the ladder of value-added products, the product development will necessarily become more sophisticated. Sub-contractors would likewise benefit from a more extensive flow of information or "specifications" and managerial strategies for production of increasingly complex products. Modular production permits isolation of codified knowledge within components which might limit transfer, and certainly the firm's concern to preserve its own "black box" technology would limit flows of proprietary knowledge. Strategic partnerships offer further channels for knowledge flows, evident already in the joint-development of mobile phones with Huawei at COSMOBIC Technology, or the new venture with Beijing Tsinghua Tellhow Intelligence Technology in energy-efficient building engineering.

What local factors might encourage foreign firms to upgrade knowledge flows on the mainland? Continuing development of the national innovation system would certainly be a factor, but in China critical factors would be both improvement in research infrastructure and access. A new emphasis on "indigenous innovation" at a recent Party Congress has caused some anxiety among foreign firms over exclusion. Added to the anxiety is the national focus on "hardware" or buildings at the national research institutes, rather than on the personnel needed to expand the system (OECD/MOST-C 2007: 26). A shift from a focus on market to a focus on China's human resources among major foreign investors may also drive individual firms to improve their knowledge flows at production sites. High-technology sub-sectors such as information and communication technology groups appear to be leading the shift and making the investments, rather than the broader range of foreign firms in other major manufacturing sectors.

Panasonic has harnessed research investment to market demand. Beijing Labs focus on higher-value-added sub-sectors, just as the Shanghai Center develops household electronics. The firm also can rely on the network of Japanese SMEs in China, important initially for reliable supply and cooperation with local [illegible] As localization proceeds, that Japanese SME network may serve as a channel for improving process and for specialized product development. The capability of bridging modular and integral architecture of electronics merchandise will likewise serve the firm well in China. METI recommended a dual strategy in the sector of automation in the machinery and process for production of modular products, and the Japanese system of cooperation within and among firms for integral products (METI 2007d). Panasonic appears well positioned to flourish in both areas in China, churning out integral architecture-based products for the upper end of the market, such as compact consumer electronics, while improving the process and automation for general products as the domestic market demands.

Some remain pessimistic over China's profitable but perhaps shallow anchor in GVCs (Lemoine and Ūnal-Kesenci 2004; Ohashi 2006). The technology regime of the electronics sector in an export-oriented development process often monopolizes the resources of local entrepreneurs and state policy-makers, due to the quick return on investment and the ready access to global chains. If indeed China has succeeded at modular architectural production in electronics and textiles at the expense of the more incremental, embedded craft and technology needed for integral product architecture, Panasonic's advantage in both modular and integral production may permit a firm-specific node with greater depth in GVCs.

Thailand: complementarity without competition

Thailand's annual production of electronics in 2007 reached $24.5 billion, up from a total of only $19 billion just three years earlier. Data-processing equipment (44 percent) and components (32 percent) represented the major areas of production. The production of color TVs had already jumped from 20 million units in 2004 to 24 million units in 2005 (Reed 2007: 202). Higher-value-added products drove growth in 2007. Exports of hard disk drives saw an annual increase of nearly 40 percent in 2007 up to $9.5 billion, as Seagate, Western Digital, Fujitsu and NEC all added capacity (Oizumi 2008a). Beyond products and production facilities, preferential trade agreements with ASEAN and India and other nations have helped spur exports of consumer electronics as well.

Electronics provide better than one-quarter of total Thai global trade. Sector exports of $130.5 billion in 2006 represented 29 percent of total Thai exports for the year (Table 5.1). How important is trade in the Thai economy? The value of Thai merchandise trade represents a remarkable 126% of GDP, reflecting extensive dependence on global trade for national economic growth. South Korea is a major trading nation, but registered a ratio of merchandise trade to GDP of only 71 percent in 2006, and China a ratio of only 67 percent (World Bank 2008). How important is the electronics sector in the Thai economy?

Manufacturing employed only 16 percent of the total workforce in a largely agrarian economy, but generated 45 percent of the value added for the national GDP in 2006 (NSOT 2007; World Bank 2008). Electronics alone registered a work force of about 370,000 in 2005, 9 percent of all workers in the nation's manufacturing sector (DTC 2007). Autos, electronics, and textiles all contribute to value-added production in manufacturing, but as the leading government technology office put it simply, "electronics and electronic parts exports are major revenue earners for the Thai economy (NSTDA 2008)."

Electronics and electrical machinery contribute greatly to the nation's technology trade as the leading source of "high technology exports," particularly in a country with very weak performance in patenting activity. A 27 percent share of "high-technology exports" in total manufactured exports in 2006 seems remarkable given that other indicators of national technology levels cited in Tables 4.1 and 4.2 above were relatively low. The dominant role of foreign firms in the sector and in trade appears to offer the explanation, and a different picture of anchoring in GVCs. Concerns raised about the shallow roots of China's electronics sector with modular production or simply assembly by foreign firms appear equally relevant for the Thai electronics sector (Lemoine and Ūnal-Kesenci 2004; Steinfeld 2004).

What I term "cooperation without local competition" is a further feature of the

Thai electronics sector. Exports to Japan of $5.2 billion represented nearly 14 percent of the total $31.5 billion. On the import side, Thailand brought in $31.6 billion of sector goods, representing 24 percent of all imports. Japan absorbed about 14 percent of the exports, but alone provided better than one-fifth of total Thai imports. Unlike Japan's trade with China and South Korea, Japan was a major importer of Thai electronics products, as well as an exporter, providing evidence of closer links between Japanese and Thai networks than in the case of Japanese and Chinese or Japanese and South Korean networks. Japan has long provided government aid to align Thai policy on SMEs, trade associations, and investment policies with Japan's own social system of production. Complementarity is the key to effective "supporting industries."

Sector

Electronics has developed as a typical export-oriented industrialization (EOI) sector in Thailand, but without comparably successful import-substitution industrialization (ISI) strategies in other upstream sectors. The result is a supporting specialized sector of local firms supplying major foreign subsidiaries on Thai soil. Sectors such as petrochemicals and steel, for example, have emerged with limited capacity and only basic products. What complements electronics in the local economy, however, is an extensive automotive industry well along the path of automotive electronics.

Beyond sectors, the corporate structure has likewise been influenced by EOI strategies. EOI usually includes majority foreign participation in joint-ventures at the outset, or full ownership within export zones. Thai Board of Investment incentives to attract, hold, and expand foreign capital in electronics, plus decades of operation by foreign firms now result in a dominant pattern of wholly owned subsidiaries rather than joint-ventures. Some have deep roots. Thailand was the site of Panasonic's very first offshore manufacturing site, established already in 1963. Thai electronics has evolved into a specialized sector with a distinctive blend of local and foreign corporate roles quite distinct from its northern neighbors in Asia.

Operating in the shadow of China and South Korea, a Thai export-oriented strategy of industrialization in electronics has brought growth to supporting SMEs, but without larger local firms to compete with foreign investors. Many developing economies without the benefit of links to global GVCs would envy the Thai network of SMEs supporting foreign firms in the sector, and the opportunities for employment, anchoring of foreign capital, and technology transfer. As the Danish Embassy in Bangkok recently reported, "Thailand has become a major manufacturing and assembly base for a number of electronic and related components, including integrated circuits, printed circuit boards, miniature ball bearings, computer cards and cables (DTC 2007)." Local SMEs provide basic inputs for manufacturing, and local engineers operate those assemblies. But how secure is that anchor?

The dominant role of foreign firms is a second feature of the sector, permitting extensive flows within GVCs, but a limited role for indigenous firms. Larger electronics firms dominate the Bangkok chapter of the Japanese Chamber of Commerce and Industry with 163 sector firms as members (JCCI-B 2007a). Nikkei affiliates in electric machinery employed about 140,000 Thai workers in 2005 according to the Toyo Keizai Databases. METI estimates that sales by Japanese affiliates up to 2003 in Thailand amounted to about $8 billion annually (Ramstatetter 2007).

A closer look at Thai trade in computers and electrical machinery highlights the extent of trade dependence among foreign subsidiaries in the Thai electronics sector (Table 5.4). At first glance the industry appeared to flourish with computer exports of $10.8 billion, compared with imports of only $1.9 billion. A different picture is apparent in the component trade, where imports of $4.2 billion outpaced exports of $3.7 billion in the flow of parts for computer and other office machines. In the broader category of electrical machinery and parts, Thai imports of $25.2 billion again outpaced exports of only $23.2 billion. In contrast, Japan, China, and South Korea achieved strong trade surpluses in all three categories.

Thai observers complain of relatively weak rooting of parts production in local SMEs. For instance Apichai Boontherawara, President of the state-owned Exim Thailand Bank, warned of the illusion of export growth in the category of high-technology goods in 2007. He explained that "80 per cent of high-tech exporters are foreign enterprises registered in Thailand" only marginally affected by currency fluctuation "due to the high import content of their export items (BP 2007b)." Recent devaluation of the US dollar against the Thai baht has eroded the competitiveness of local firms in global GVCs.

"Transition" is a third feature of the sector, including changes in trade ties as

Table 5.4 Thailand's world trade in electronics*, and Japan share, 2006 ($US millions)

Trade source/ destination	*Value of data-processing machines (Code 8471)*	*Value of parts for data-processing machines (Code 8473)*	*Value of electrical machinery and parts (Code 85)*	*Value of electronic merchandise (Codes 8469–8473, 85)***
Imports from Japan	*87.1*	*270.7*	*6,821.9*	*6,648.8*
Total imports	1,920.7	4,295.4	25,248.8	31,592.5
Japan share	4%	6%	27%	21%
Exports to Japan	*856.5*	*88.0*	*4,303.9*	*5,249.6*
Total exports	10,849.6	3,761.4	23,256.0	37,875.8
Japan share	7.8%	2%	18.5%	13.8%

Source: UN (2008).

* HS 2002 SITC Codes: Code 8471, automatic data-processing machines including computers; Code 8473, parts and accessories for use in computers and other office machinery; and Code 85, electrical machinery and equipment and parts thereof.

** Includes the following HS 2002 SITC Codes: Code 8471, automatic data-processing machines including computers; Code 8469, electric typewriters; Code 8470, calculating machines; Code 8471, automatic data-processing; Code 8472, other office machines; and Code 8473, parts and accessories thereof, plus Category 85, electrical machinery and equipment and parts thereof.

well as technology. The technical level of Thai supporting industries in modular production has won recognition from surveys of Japanese investors in Thailand, often attributed to Japanese technical support and cooperation with Japanese SMEs in Thailand. Local SMEs appear to be a successful case of the growth among supporting industries. Indeed, the transition from simply appliances to hard disk drives and semi-conductors indicates how far the nation's electronics sector has moved up the value chain. In addition, the recently concluded Japan–Thailand Economic Partnership Agreement (JTEPA) provides extensive institutional support for cooperation between SMEs of the two nations, and for technical cooperation in specific sectors. Beyond JTEPA, Thailand's successful pursuit of FTAs with China, India, Australia and other countries, in addition to trade liberalization within ASEAN, has enhanced Thailand's status as a major electronics hub in Southeast Asia. Japanese firms can now coordinate production and marketing through the hub.

Panasonic

Recent examples of re-positioning to take advantage of trade pacts include Panasonic's relocation of ventilation fan manufacturing from China to Thailand. Daizo Ito is the CEO of the Panasonic Group in Thailand. He explained, "We decided to promote Thailand as a production centre for certain products in the region to take advantage of ASEAN free-trade regulations that allow us to export products at tariff rates of between zero and five per cent (TN 2006a)." A recent JETRO survey of Japanese manufacturers in ASEAN found that not only do they export much of their production, but that 40 percent of Japanese firms in Thailand import at least half of their raw materials and parts with no tariff whatsoever (JETRO 2007a).

CEO Ito enumerated twenty-three separate businesses and fifteen manufacturing plants in 2006, fourteen of which produce for export. Products include washing machines, refrigerators and air-conditioning units, as well as electric components and parts (TN 2006a). The manufacturing subsidiaries listed in 2007 employed over 7,000 workers (TKS 2007a). Ito spoke the next year of vulnerability to the baht appreciation against the dollar, since almost 70 percent of Thai Panasonic sales of $1.9 billion in fiscal 2007 came from exports (TN 2007c).

Prospect for knowledge networks

Thailand's electronics and electric machinery sector alerts us to the significance of supporting industries for knowledge networks. Foreign firms play a major role in electronics in both Thailand and China, but without competition from major local firms in Thailand, and without the attraction of a domestic market similar to China's. The absence of both competitive larger firms in the electronics sector, or of joint-ventures with local firms precludes alliances in R&D evident in China's and South Korea's electronics sector. Extensive networks of Japanese subsidiaries in Thailand, however, and of their local suppliers, provide opportunities for knowledge flows quite different from the South Korean sectoral landscape. Japan and

Thailand agreed to foster close cooperation between SMEs of the two nations in Chapter 11 of the JTEPA "Implementation" document (JTEPO 2007). Yet somewhat surprisingly, there was a commitment to facilitate investment of Japanese SMEs in Thailand for the express purpose of growing business in Southeast Asia – a regional rather than a bilateral strategy (Article 50). Building upon decades of investment and sub-contracting in the Thai sector, JTEPA provides new institutional support for knowledge flows in Thai "supporting industries."

Electronics figured prominently from the very start of JTEPA negotiations with the initial task force in 2003. Among the four corporate executives from Japan's leading business association chosen to represent Japan, executives from Toshiba and Omron represented electronics, and leaders from Teijin and Toray represented textiles (EPA 2003). The Task Force recognized Japanese efforts to develop Thai SMEs from the early 1990s, and recommended dispatch of still more Japanese SME experts to Thailand to provide technical training. Japan's International Cooperation Agency that year had noted that Japan was Thailand's leading trade partner, foreign investor, and bilateral donor. Grant aid concluded in 1999, but technical cooperation and yen loans had been expanded. The report included a critical recommendation based on long cultivation of complementarities: "Japan should appreciate personal and well-organized networks that have been developed between the two countries over the years through development assistance and put them to good use for problem-solving (JICA 2003b: 26)."

Thai leaders also found benefit in the partnership, particularly as the agreement moved closer to ratification. Thai Commerce Minister Somkid envisioned Japan as the "kingdom's primary business partner in developing Thailand as an Asian hub for electronics and electrical equipment production (MCOT 2006)." He cited "partnership" rather than trade expansion as the main goal, including Japanese "collaboration in technology and human resource development." As Somkid emphasized, "Japanese technology will help Thailand maximize its business potential." Thailand would appear to fit the Japanese model of a supporting industry offshore, with technical guidance of Nikkei SMEs and their experts, a focus on craftsmanship (*monozukuri*) in manufacturing, but mainly training and technology transfer rather than R&D. One finds no reference in the JTEPA documents for "cooperative learning" such as specialized research or product development evident among supporting industries in Japan.

Growth in trade and FDI provide one indicator of business potential, highlighted in investor confidence evident in recent surveys. As for technology levels, a majority of Japanese firms with manufacturing investment abroad in East Asia agree that general purpose components from ASEAN countries for the manufacture of electrical machinery appear comparable now to technology levels in Japan (METI 2007d). On the Thai side, electronics merit a high research priority in the national innovation system, focused on linking government-funded institutes to industry (NSTDA 2008). President Charuek of the leading sectoral association spoke publicly of the transition from commodity to higher-value-added products as the only path of Thai survival (TN 2007d). His group works to consolidate gains in order to anchor investment in the intense global competition for upgrading industries.

They take as their focus standards and certifications, as well as capacity-building in domestic SMEs (EEI 2007).

Singapore and Malaysia have long captured the higher-end research in electronics for the region. But the Thai transition from home appliances to semi-conductors has increased production at home of higher-value-added goods. On the consumption side, this demand for more sophisticated consumer electronics will draw more market research to the country for product development. Panasonic opened its Home Appliances R&D Center for the Asian Region in November of 2005 on an industrial estate east of Bangkok. Tasks include planning and development of refrigerators, washing machines, etc. (TKS 2007a). Booming auto exports may draw further attention from electronics firms on product development. Panasonic has already moved part of its R&D work on car radio and navigation systems to a new center in Bangplee, Samut Prakan outside of Bangkok, the home of both Toyota and Denso in Thailand (TN 2006b).

Conclusion

A review of Japan's sectoral networks in East Asian electronics and electrical machinery draws us from profile to prospect. Two features distinguish the South Korean sector from comparable sectors in China and Thailand. One is the relatively limited presence of major Japanese manufacturing subsidiaries on the peninsula. A second feature is the extensive structural compatibility between the sectors in Korea and Japan, with concentration in major local firms, localization at the expense of discouraging inward FDI, and strong technology regimes for the sector in both countries. A profile of the South Korean sector highlighted strong local firms and contributing local SMEs, an asymmetric interdependence on Japanese components, and a growing number of strategic alliances with foreign firms. The latter provides openings for further cooperation in development of advanced technology for specialized product areas.

The challenge now is to develop links among SMEs and a mutually beneficial division of labor in innovation between the two nations. Knowledge networks in supporting industries through cooperation among independent contracting SMEs in both countries have already been cited as a possible benefit of the now dormant proposal for a Japan–Korea FTA. South Korea is already the third largest holder of patents in the Japanese market (MEXT 2007: 150). The study reminds us that complementarity in social systems of production by itself does not necessarily lead to collaboration. Extensive compatibilities of industry association, standards, and incremental learning between the sectors of the two nations can also contribute to intense rivalries in exclusive knowledge flows.

A growing number of in-house research units together with the manufacture of more sophisticated products offers hope for better knowledge flows at Nikkei subsidiaries in China. A shift from being merely the world's factory for electronic merchandise to being one of the world's leading makers and *markets* was one feature of the sector. Apart from an expanding consumer base at home, rapid technological development among indigenous firms and extensive foreign investment drive the sector at home and abroad. Prospects for more extensive knowledge networks in product development appear strong, but not necessarily in advanced research at this point. Moreover, electronics may not be representative of the wider industrial base. Modular production in the electronics sector may actually discourage the more extensive knowledge flows familiar in integral production. China's electronic sector remains an experiment in rapid development through extensive engagement with GVCs. The dominant role of subsidiaries of foreign firms in the "high-technology" trade and product development for the local market may spur new knowledge flows, but may not necessarily benefit local firms or contribute to the national innovation system.

Electronics in Thailand prosper with neither the dominant local firms evident in South Korea, nor the blend of diversified foreign firms and local competitors evident in China. We find rather a specialized, supporting sector, with aspirations as the Asian hub for the Japanese electronics industry. A transition in the sector from electrical machinery to electronics suggests that EOI strategies cannot proceed on their own plane, independent of deep local roots or densely embedded local networks. The transition offers new hope for knowledge networks in Thai supporting industries, though limited to basic production tasks, and relying in part on the new partnership agreement with Japan.

Appreciation of the Thai currency against the US dollar in 2007 exposed the vulnerability for local players in a foreign-invested industry. Less dependent on local inputs or the local currency, foreign subsidiaries can flourish with their imports of components and exports of finished goods. Trade volumes and value grow, but local firms in the sector may suffer. A further lesson evident in Thai electronics is the fragility of innovation drivers. Demand from foreign markets and makers appears to have driven a Thai transition from appliances to semi-conductors, with adequate local engineering and basic labor resources only a contributing factor. Local drivers would be critical for the long-term anchoring of innovation.

Boundaries

States face hard choices in the effort to balance regional with national interest, integration and insulation. The term *boundary* offers a useful focus for tracking the shift from assembly to manufacture within more intensive knowledge networks. States try to gain entry at the boundaries of GVCs with incentives for IFDI harnessed to import substitution and/or export promotion. Further regional integration may erode state efforts to insulate national innovation networks, yet the state remains a critical player in the competition for IFDI to move industry up the value chain. States which set boundaries for national innovation systems also manage the task of improving innovation systems. But even here, borders appear increasingly porous. For instance, the South Korean government attracts FDI by offering foreign firms access to government-supported research programs. State oversight often gives way to corporate interests with the growing complexity of boundaries of investment, as evident in technology regimes of specific sectors. The prevalence of research institutes within firms rather than in cooperative networks among university, industry, and state offers further evidence of the rise of corporate interests in the innovation chain (Hershberg, *et al.* 2007).

Beyond boundaries of GVCs and state innovation systems, corporate boundaries take center stage in discussion of strategic alliances, subsidiaries, and contractors within supporting industries, or of alliances with states in techno-nationalist or meta-national strategies. The changing market boundaries evident in ASEAN can bring new networks or strengthen manufacturing networks with product development closer to consumers. Preferential trade agreements have begun to change the role of national boundaries in corporate knowledge networks, permitting horizontal integration for foreign-invested corporations on site abroad. Liberalization of trade flows has permitted new knowledge flows, often through a triangular strategy of foreign subsidiary sourcing in the region from a Thai base, and then exporting production from the Thai base.

Networks, permeable or insulated, still have boundaries. High technology in the electronics industry has forced firms to seek partnerships due to the rapid pace of technological advances and the challenges of marketing new products. How do Japanese firms maintain boundaries in the face of growing inter-penetration of firms in product development? One scholar has done extensive research on the restructuring of Panasonic through a specific type of network called "strategic communities (Kodama 2007)." Strategies differed with products and participants, but the "embedded" character of the communities remained a constant. The term denotes a critical feature of integration and communication whereby different strategic groups can share knowledge within the organization beyond the boundaries of their own unit (Kodama 2005). Sharing tacit and explicit knowledge across the organization beyond individual units of marketing or product design was the key to generating new knowledge relevant to both consumer and to the technology resources of the firm.

Kodama provided ample evidence of the effort within a vertically integrated business model at Panasonic. He highlighted, for instance, data on embedded communities cooperating for marketing in a horizontally integrated business model. He then suggested, but did not provide convincing evidence, that high-technology firms used the same strategy in alliances to develop technology platforms through the sharing of core technologies or OEM production, though not specifically at Panasonic. Nonetheless, the thesis of embedded strategic communities for innovation at Panasonic reflects the high context character of Japanese innovation, further evidence that Japan is indeed a network society (Kumon 1992). The thesis also reinforces the effort at "partnerships" with other countries, institutionalizing compatibilities for the long term, and cooperating in the development of supporting industries. Finally, the thesis adds organizational detail to Lincoln's emphasis on knowledge sharing in the firm's "cooperatives" (*kyōei-kai*) of sub-contractors in domestic supporting industries (Lincoln 2006).

6 Automotive sector

Global models and local minds

Analysis of a domestic auto industry opens a window on the broader structure of industry at home. A current profile of change in the sector draws attention to major global assemblers in regional networks, and a new division of labor in the flow of parts across borders. I begin with local sectors in each country that might promote or discourage knowledge exchange. I then track Nikkei networks in each country to assess precedent and prospect for knowledge networks. South Korea, China, and Thailand have strong sectors and deep ties to global value chains in auto production, but with widely discrepant anchors in the value chains of Japanese autos. Can the target nations now secure innovation?

If significant for tracking social systems of production, the auto sector remains largely opaque. The embedded character of such a complex industry complicates the task of tracking value chains in the sector, in contrast to analysis of more standardized supply chains in electronics textiles and garments. Nonetheless autos, electronics, and textile industries worldwide face a similar challenge of anchoring production of global models, and capturing a place for local minds in innovation efforts.

Just as electronics often leads export-oriented strategies of development, the auto industry drives import substitution. Automobiles can stimulate domestic growth, pulling along a variety of sectors with extensive and sustained demand. Parts in a single vehicle may total up to 30,000 separate elements crafted out of materials as diverse as rubber, plastic, metals, textiles, and glass (JAMA 2007a). Electrification of critical controls such as fuel injection, steering and breaking, has eroded sectoral boundaries between autos and electronics whether at a global supplier such as Denso, or at an electronics giant such as Panasonic. Successful modularization in electronics has inspired efforts to standardize modules in component production, but with limited success. Joining local minds to global models in integral rather than modular manufacture in autos presents opportunities and challenges quite distinct from those present in the electronics sector.

Nikkei networks

Japan competes in South Korea, cooperates in China, and commands in Thailand. Autos are paramount among Nikkei investment networks in East Asia, given the prominence of both Japan and the region in global auto production, and Japan's deep regional ties in manufacture. Japan, China, and South Korea ranked among the top five global auto-producing nations in 2007. Beyond simply production, the nations rank among the world's top exporters of vehicles. Japan perennially captures the top spot, with South Korea not far behind in fourth place (Table 6.1). Thailand has emerged as the leading auto production center among the ASEAN nations, and a fast-growing base for auto exports. Apart from global prominence in manufacture and trade, the four nations have also developed extensive auto parts industries to serve both local and regional producers. Finally, leading auto assemblers worldwide have invested in research centers in all four nations, whether for product adaptation to local regulations and market demand, or for engineering platforms for export production.

The automotive sector demonstrates differences and highlights dynamics of corporate scale in Nikkei networks, shedding a powerful light on ties between smaller and larger firms, and particularly in patterns of cooperation among the second, third, and fourth tier of suppliers. Japan's model of cooperative learning and what Dore termed "relational contracting" developed initially in the auto industry as assemblers relied on independent firms for parts and component manufacture, in contrast to the Detroit model of vertical integration. Integrated manufacturing in the auto sector provides a foil to modular production prominent in electronics, stretching analysis of knowledge networks deeper into the fabric of multiple suppliers cooperating on specifications in the absence of standardized interfaces.

Nikkei chains of manufacturing and supply differ dramatically in the three nations. I look to those differences in this chapter to learn of knowledge exchange in the three target nations at Toyota and Denso affiliates. Toyota reported a net income of $13.9 billion for fiscal 2007, with a global research budget of $7.5 billion (Ward's 2008a: 83). Denso reported a net income of $1.7 billion for the same period, with a research budget of $2.4 billion (Denso 2008b). Toyota creates and adapts models and modules, while Denso creates and adapts products and production systems. Nikkei networks in each country bring similar specifications for global models, but with far different local contributions to global modeling. Differential embedding of Nikkei automotive networks in the three nations highlights the current diversity and future prospect of knowledge flows for region and nation.

How then do Japanese auto firms move abroad with their model of cooperative learning to integrate local suppliers into their value chains? The industry has been a bellwether for tracking "transfer of technology" between Japan and her neighbors. We find on the one hand, a familiar pattern of keeping research at home while sending production abroad. On the other hand, Toyota's research and development centers in Tianjin and Bangkok may augur well for new research links in the knowledge societies of East Asia.

South Korea: complementarity without cooperation

South Korea has followed the development model of Japan, the USA, France and Germany with a "full-set" industrial base including steel, petrochemicals, electronics, machinery, tool and die sectors. Concentration of capital and automotive technology within major firms followed the *jaebeol*, or diversified family-owned conglomerate pattern in South Korea, paralleling the earlier diversified Japanese *zaibatsu*. Multiple tiers of independent suppliers and sub-assemblers serving *jaebeol*-dominated firms such as Hyundai, Ssangyong and Samsung is again reminiscent of the Japanese industry, but with curiously weak ties to Japan. If neither Toyota nor Nissan manufactures on the peninsula, what would drive knowledge exchange in auto production between the two nations? Can a nation anchor innovation without first anchoring production? I begin with the firms, continue with their partners and auto parts, then conclude with profile and prospect for knowledge exchange.

The nation takes pride in its own global models. Hyundai–Kia and other South Korean auto firms had gained fifth place among the world's largest auto producers already by the mid-1990s, and weathered the Financial Crisis to recover earlier production levels by 2000. South Korean automakers churned out over four million vehicles in 2007, quite an achievement for the world's eleventh largest economy. Perhaps more remarkable is that fact that the firms exported over half of those vehicles (Table 6.1). Export totals appear still more surprising when coupled with the fact of minimal imports. Evolution of the local industry reversed the conventional progression from saturated local market to exports, or from import-substitution to export-oriented industrialization.

Unlike most national auto industries, the South Korean industry was organized initially in the early 1970s to serve foreign markets rather than to substitute for imports locally, and then grew rapidly with exports to the USA. The South Korean state began to encourage local demand only in the past two decades, and limited the local market to vehicles made in Korea. Registrations reached the plateau of one passenger car for every four people in a population of 46.8 million people by 1999, with sales continuing to expand through 2007. State constraints on foreign

Table 6.1 Auto industry in East Asia, 2007

Auto industry/Nation	*Japan*	*China*	*South Korea*	*Thailand*
Registrations 2006 (population per vehicle)	74.2 million (127,515/ vehicle)	35 million (2%)	15.8 million (34%)	9.5 million (64,632/ vehicle)
Units produced 2007 (world rank)	11,596,327 (#1)	8,882,456 (#3)	4,086,308 (#5)	1,125,316
Vehicle exports 2007 (world rank)	6.5 million (#1)		2.8 million (#4)	690,100*

Source: Ward's (2008a).

* Source for Thai vehicle exports: TAI (2008).

direct investment (FDI) in the assembly sector coincided with the sector's general policy of localization of industry.

A snapshot of the automotive trade in 2007 highlights contributions to a trade surplus. Imports of parts as well as vehicles amounted to $5.2 billion, just less than 2 percent of the total annual value of imports. The nation registered $42.6 billion in automotive exports, 13 percent of the total value of South Korean exports (Table 6.2). Few industries better convey the new competition for knowledge in East Asia. A national day to celebrate achievements in the industry, including growing exports as well as offshore production, provides ample evidence of sectoral prominence in South Korea's economy, and a long precedent of state–industry cooperation.

Leading the dignitaries at South Korea's National Automotive Day in May 2008 was none other than the Knowledge Economy Vice-Minister, Lim Chae-min (KT 2008c). Minister Lim's presence at the event left little doubt about sustained state priorities on knowledge creation for the global competition in the auto industry. National champions like Hyundai–Kia among the assemblers dominate local sales and local production. Their parts affiliate, Hyundai Mobis, plays a similar role in scale of production in the local parts industry, though not in technology leadership. Will South Korean and Japanese firms find common ground for knowledge exchange in the global market?

Sector

What might explain the absence of Japanese majors in South Korea's auto industry? State policy largely excluded foreign competitors from auto assembly on Korean soil until the past decade. Renault of France acquired a South Korean firm in 2000, and joined it to their alliance partner, Nissan of Japan, for marketing but not production. General Motors (GM) of the USA followed in 2002, with investment partners such as Suzuki of Japan and Shanghai Automotive Industry Corporation (SAIC) of China. SAIC then acquired a separate venture in South

Table 6.2 Value of South Korea's trade in vehicles and vehicle parts, 2006 ($US millions)

Partner/ Commodity	*Automotive sector trade (A)**	*Parts sub-sector trade (B)***	*Parts/Vehicle (A/B)*	*Total Korean commodity trade (C)*	*Automotive/ Total (A/C)*
Imports	5,242.0	2,549.0	48.6%	309,379.4	1.69%
Export	42,605.2	9,498.9	22.3%	325,457.2	13.09%

Source: UN (2008).

* HS 2002 SITC Code 87, vehicles other than railway or tramway rolling stock. This is the broader sectoral category, including vehicles and parts.

** Combination of HS 2002 SITC Sub-codes 8706–8708. Code 8706 includes chassis fitted with engines, Code 8707 includes bodies for motor vehicles, and Code 8708 includes parts and accessories of motor vehicles.

Korea in 2004. Knowledge flows within the South Korean industry today include closer integration into global flows with major Western firms on its soil, and more two-way flows through research units of foreign firms on the peninsula, but weak ties with neighboring Japanese giants, Toyota and Nissan.

Local firms have thrived in global knowledge streams. Flows include acquisition of technology from foreign firms, extension of South Korean firms into foreign countries for market access and production facilities, and cooperation with counterpart auto industry business associations abroad and their dialogues within APEC on trade and standards. Perhaps most significant are sustained the ties with foreign firms on Korean soil in manufacture and supply of auto parts, a major contrast with the assembly side, and also with the trajectory of the Japanese auto industry. South Korea remains dependent on foreign supply of certain high-value-added auto parts such as sophisticated fuel injection systems, despite rapid development of sophisticated auto design capabilities.

Western partners

Unlike South Korea's electronics sector, its auto industry has been much affected by FDI. Mitsubishi Motors and its partners retain only a 5 percent share in Hyundai Motors, although Mitsubishi provided much of the initial technology in the firm's development. Hyundai bought out Kia motors at the turn of the century, and now ranks seventh among the leading auto firms with production in 2007 of 3.7 million units. Daewoo Motor was originally a joint venture with General Motors, but bought the foreign share in 1992. The next decade brought a dramatic reversal. With Daewoo struggling on its own under the weight of foreign debts and over-extension abroad, General Motors and its partners reacquired the firm's main plants in South Korea and Vietnam in 2002 (McNamara 2005).

GM increased its holdings to a 50 percent ownership share by 2008, with affiliate Suzuki of Japan holding 11 percent, and partner SAIC of China holding 10 percent. "GM Daewoo" churned out 1.8 million units in 2007, nearly half the total production of its local competitor, Hyundai–Kia. GM quickly turned local design and engineering to its advantage with manufacture of the Lacetti Compact based on a Daewoo design. The firm soon organized global small and mini-car architectural development teams at the GM Daewoo Design Center and GM Daewoo Technical Center in Bupyeong, South Korea (GM 2008).

Two other smaller joint-ventures with foreign producers have likewise helped to change the landscape of local ownership. SAIC itself acquired a smaller South Korean producer, Ssangyong Motors, in October 2004 (Ssangyong 2008). A domestic Chinese firm now has its own independent foothold in the South Korean industry, in addition to partnering with GM Daewoo. Most global assemblers have production facilities in China, but few have facilities in each other's territory. Moreover, SAIC has moved into the manufacturing side of the Korean industry in tandem with its Western partner, GM, but before Japanese firms have moved manufacturing to South Korea.

A second venture indicates as well the major role of Western firms on the peninsula in the absence of independent Japanese competitors. Samsung secured technical support from Nissan of Japan in the early 1990s to launch auto manufacture. Renault subsequently obtained an 80 percent share of the Samsung auto in 2000, and reorganized the venture under the Renault Nissan umbrella. Today Renault Samsung manufactures three models based on Nissan designs, and operates a Technical Center and a Design Engineering Center (Renault 2008). Apart from this mediated relationship with Nissan, Japan remains conspicuously absent from the new landscape today of diversified foreign investment in assembly on the peninsula. In retrospect, the South Korean automobile industry was remarkably international from the outset with respect to ambitions, markets and technology, yet remarkably local in its ownership until recently.

Auto parts

One major difference in South Korean auto manufacture versus that in Japan and the West has been the late development of the Korean auto parts sector. Lacking the sophistication and scope of products to meet the demand of the auto manufacturers, South Korean assemblers remain dependent on foreign auto parts firms for critical components. Imports of nearly $1 billion worth of parts from Japan in 2007 stood in stark contrast to exports of only $368 million in parts (Table 6.3). But South Korea's asymmetric interdependence with Japanese auto parts suppliers differs significantly from the reliance noted above in electronics. One difference is the extension of a variety of foreign auto parts firms to the peninsula, including US firms. State efforts to sustain the industry in the early decades included efforts to insure local ownership in assembly, but development of the complex supply lines of components and parts remained closely linked to investment and production by mainly Japanese suppliers.

Table 6.3 Value of South Korea's trade in vehicles* and vehicle parts** with China and Japan, 2006 ($US millions)

Partner/Commodity	*Vehicle sector trade (share of total Korea trade) (A)*	*Parts sub-sector trade (B)*	*Vehicle/Parts (A/B)*
Import from China	457.3 (8.72%)	285.8	62.5%
Export to China	3,153.0 (7.40%)	2,551.9	80.5%
Import from Japan	1,429.9 (0.46%)	933.2	65.3%
Export to Japan	402.0 (0.12%)	368.0	25.7%

Source: UN (2008).

* HS 2002 SITC Code 87, vehicles other than railway or tramway rolling stock. This is the broader sectoral category, including vehicles and parts.

** Combination of HS 2002 SITC Sub-codes 8706–8708. Code 8706 includes chassis fitted with engines, Code 8707 includes bodies for motor vehicles, and Code 8708 includes parts and accessories of motor vehicles.

A second difference is the strong role of global auto assemblers on the peninsula, whether in the origins of the industry or in the recent restructuring, diversifying interdependence. Officials of the Trade Promotion Organization reported that local assemblers drew 35 percent of the total value of parts and components in 2003 from foreign-invested firms on the peninsula. Of nearly 900 firms registered in the South Korean auto parts sub-sector in that year, 227 foreign-invested firms accounted for one-fifth of the total. For instance, Delphi employed nearly 4,300 in six joint-ventures and one subsidiary by 2003, as well as a research institute in Seoul for systems engineering. Company officials at the time cited proximity to research institutes of assemblers as the advantage of a Seoul location (KOTRA 2008b; 2007). Despite the firm's recent financial problems, the corporation has maintained its South Korea network (Delphi 2003; 2008).

Denso

Unlike electronics, auto production demands a broad variety of crafted components ranging from simple bolts and screws to sophisticated control systems. Japanese policy-makers as well as firms like Denso take pride in *monozukuri*, the "art of making things (Denso 2008b)." South Korean sub-contractors could benefit from a long Denso tradition of sustained improvement of products and production systems. Technology levels and investments of Korean SMEs would appear ideal partners for some variation of cooperative learning. Baek and Jones noted the prominent share of overall research spending at Korean SMEs versus larger firms (2005). METI officials themselves have recognized Japan's interest in improving the technological abilities of affiliated Asian SMEs for commodity-type production (METI 2006a).

Denso enjoys a rare pedigree in South Korea with technical ties dating from 1969 when Denso and another Toyota family supplier, Asmo, supported the early development of one of Korea's larger suppliers of fuel pumps and electronic parts, Poong Sung Electric. Denso and Asmo developed two production bases on the peninsula, a joint-venture with Poong Sung titled Poong Sung Precision Company from 1976 for the manufacturing of an instrument cluster. Investment continued a decade later with Korea Wiper Blade Company, and then a sales office in 1997. The two Toyota partners joined in a new venture with Poong Sung from 2000, titled Denso PS, with Denso as the leading shareholder (Denso 2000). Employment had grown to about 1,400 by 2008, with products such as small motors and electrical parts.

In line with their stated "Vision," Denso's South Korean investments would localize business operations with the goal of higher reliability and lower costs. Moreover, the website for global operations highlights local contributions: "Denso harnesses the worldwide strength of our companies to bring locally developed technologies and know-how to new products. For example, our car navigation systems for customers worldwide feature core technologies created in Japan, Germany and China (Denso 2008a)." Why not in South Korea? Denso appears to be growing outward rather than inward in Korea. The firm has taken a new direction on the peninsula recently in off-shoring production from a Korean base.

Following expansion of Hyundai–Kia into Beijing, Denso allied with Denso PS to manufacture instrument clusters at a facility in nearby Tianjin, "Tianjin Poong Sung Electronics Company (Denso 2006)." The Japanese firm thus uses its South Korean subsidiary not only to supply local assemblers on the peninsula, but now to follow Korean assemblers to China to supply them directly on the continent. But compared to Denso's production networks in China or even Thailand, production on the peninsula remains quite limited. Denso's global operations website lists four firms in South Korea with a total workforce of about 2,200. The firm lists twenty-four firms in China with a workforce of 7,500. Only eight ventures are listed in Thailand, but with a workforce of 9,400.

Denso's production facilities provide ample evidence of the Japanese role in South Korean auto parts. Some might suggest that Denso has succeeded where Toyota has not. But the past here is not necessarily prologue for foreign firms in South Korean auto production. Denso recently announced plans for a new fuel-injection plant in Gyeonggi Province near Seoul, a hub of foreign-invested plants. Denso's position permits access to networks of Japanese SMEs plus proximity to the research institutes of Hyundai–Kia where they can join in product development (KT 2007). Wedged technologically between the sophisticated Japanese auto industry and the rapidly improving engineering capability of the foreign-invested Chinese industry, South Korean firms have a strong interest in improving knowledge exchange with Japan, and Denso may play a lead role in achieving such an outcome. In the absence of a foothold in assembly on the peninsula, and the growing competitiveness of Hyundai–Kia, Toyota and others may prefer to cultivate knowledge networks in South Korea's neighbors such as China and Thailand.

Prospect for knowledge networks

What does the profile of the South Korean auto sector suggest for the future of knowledge networks between Japan and South Korea? The organization of the Korean auto industry into clusters to the south such as the Changweon complex facilitates knowledge flows around engineering, whether between larger and smaller firms, or firms of different nationalities. Clustering of foreign-invested manufacturing firms in Gyeonggi Province near Seoul offers further opportunity, especially for larger Japanese firms making use of Nikkei networks of SMEs already on the ground in Korea. But Japanese transportation firms recently rated the South Korean agglomeration of local firms and R&D institutions as less attractive than the clustering in ASEAN or China (Ito, *et al.* 2007: 32).

If similarities in social systems of production alone determined research cooperation, few nations would appear more complementary than Japan and South Korea in state–industry ties, organization of trade associations, and organization and social prominence of major auto firms. Sophistication of engineering in Korea, and the strength of the national innovation system might well attract more Japanese firms to the peninsula, but for higher-value-added production in Nikkei regional production networks beyond basic commodities. Denso plans to double the number of local engineers in South Korea and Southeast Asia from 400 to 800 by 2010 (Denso 2008a). The firm estimates that local subsidiaries in East Asia are already responsible for 40 percent of application design. They plan to increase the local ratio to 60 percent by 2010. A corps of Japanese SMEs as well as larger firms such as Denso and Yazaki with production facilities in South Korea already facilitates knowledge transfers on the shop floor with local sub-contractors through labor mobility, supervision of specifications, and shared production in joint-ventures.

Apart from plans, the placement of new research facilities near the larger research facilities of major assemblers such as Hyundai–Kia is critical. Participation in the "design-in" phase of planning and design gives participating parts firms the opportunity to both learn and adapt their engineering skills to the needs of the assemblers. Denso and Delphi offer precedent for other foreign firms to establish their place in the local industry. What distinguishes the prospects for the two firms is their primary linkage to major assemblers. Formerly a subsidiary of GM, Delphi enjoys strong ties to GM Daewoo. The absence of Toyota manufacturing plants in South Korea leaves Denso without the security of supplying its major shareholder.

Beyond the firms themselves, auto industry associations in the two nations find plenty of common ground with regional and global issues of property rights protection. The growing sophistication of the South Korean auto industry has engendered common interests in other areas such as regional standards, including environmental standards. Officials of leading associations such as the Japan Automobile Manufacturers' Association (JAMA), and Korean Automobile Manufacturers' Association (KAMA), consult regularly at major auto shows in Detroit and East Asia. Both nations participate as well in the APEC Automotive Dialogue.

Some look to new streams of Japanese FDI to the peninsula to replace South Korean imports of Japanese auto components and parts. Yamazawa pointed to this benefit in early discussions of an FTA between the two countries: Japanese exporters would move production facilities to the peninsula (2000). But others foresee far broader knowledge flows which would tap the geographical advantages of proximity. Kim Dohyung wrote grandly of a regional "automobile specialization area" of production networks joining the Kita-Kyushu and Western Honshu areas of Japan to the complexes on the southern coast of South Korea in Busan and the Changweon complex. Mutual development of the modularization of production and consolidation of platforms would be the key, with Japan competing at the high end, and South Korea with "mid-quality" cars (2007).

JETRO officials noted an industry preference in Japan for "closed modularization," with the production process mainly "in-house," versus a practice in Western firms of "open modularization" dependent on a number of outside sub-contractors (JETRO 2007a: 186). The fact underlines the importance of what Kim termed "strategic alliances and even mergers" among automakers across the two nations to encourage exchange (2007). Anchoring Japanese investment in the South Korean auto industry would appear to be a prerequisite for promoting knowledge flows, but also a formidable challenge in two nations with deep legacies of nationalist protection in domestic auto markets and industry.

China: cooperation without complementarity

Japanese automakers have moved alone only recently to set up sales offices in South Korea. But in China they have joined with local partners now for over a decade to import, manufacture and sell. Yet a third profile is apparent in Thailand. There long-tenured Nikkei makers have bought out local partners and now manufacture, sell, and export pretty much alone. State policy is one factor in the differences. An aggressive rush of Nikkei automakers to China provides a foil to absence in South Korea and autonomy in Thailand. China has harnessed foreign investment and technology to development of local firms.

A huge pool of skilled labor and a domestic market of 1.3 billion people have attracted all the global auto assembly and parts firms to the continent, including recently Hyundai–Kia and Mobis of South Korea and all the Japanese majors. But Japan's giants in China today, such as Toyota and Denso, face a difficult transition from patrons to partners with the growing technological sophistication of their joint-venture partners, and of the Chinese consumer, and state pressure for localization.

Sector

Much like South Korea and Japan, China has taken the path of "full-set" industrialization, supporting auto production with local steel, machinery, petrochemicals, electronics and textiles. Auto production organized within massive state-owned enterprises followed a conventional import-substitution strategy, though with extensive IFDI. In contrast, the South Korean state permitted technology tie-ups but discouraged foreign ownership until recently. China's path of welcoming joint-ventures with global leaders permitted rapid development but without extensive localization of the industry. As opposed to the nation's success in localizing modular production in electronics with competitive firms in consumer electronics, the craft and organization demanded in integral production for the more complex process of auto production has proved more difficult. Altenburg and colleagues argue that producing under license from foreign manufacturers, and a tradition of reverse engineering, have not resulted to date in development of innovation capability at local firms (2008).

China ranked third worldwide with production of 8.8 million vehicles in 2007, 12 percent of total global production. Local firms such as First Auto Works (FAW), Beijing Auto, Dongfeng and Cherry Auto, as well as SAIC, ranked among the top thirty largest producers (Ward's 2008b). FAW prospers in joint-ventures with Toyota and Volkswagen of Germany, while SAIC has partnered with GM and Volkswagen. FAW–Volkswagen, Shanghai Volkswagen, and Shanghai GM remain the leaders in the China market. But Chinese firms to date have not succeeded abroad on their own, serving as evidence of persisting reliance on the technology available in joint-ventures. Zhang Guobao, Vice Minister of the National Development and Reform Commission, spoke recently of three challenges in the industry: overcapacity, an underdeveloped auto parts industry, and unsatisfactory aftermarket service (Ward's 2008a: 38).

China's imports in the vehicle sector amounted to $17 billion in 2006, with over half the total spent on auto parts (Table 6.4). Imports worth $5.3 billion from Japan alone amounted to about one-third (31 percent) of total sector imports, with parts absorbing the major share (64 percent) of the total (Table 6.5). Exports to Japan in the sector of only $2 billion left China with a major trade deficit for the sector. On the export side, auto parts represented 61 percent of the total value of vehicles and parts sent abroad. We would expect much of this trade to represent intra-firm exchanges, particularly in view of the deep networks of Toyota family firms in parts and assembly on the continent. A similar intra-firm trade might be assumed on the South Korean side, reflecting the strength of both Japan and Korean industries in China, and their growing interdependence on the Chinese industry. Oddly, however, the relatively meager South Korean trade with Japan highlights the weak ties between the industries of geographically proximate neighbors (Table 6.3).

Table 6.4 Value of China's trade in vehicles and vehicle parts, 2006 ($US millions)

Partner/ Commodity	*Vehicle sector trade (A)**	*Parts sub-sector trade (B)***	*Parts/Vehicle (A/B)*	*Total China commodity trade (C)*	*Vehicle/Total (A/C)*
Imports	17,051.7	9,042.5	53.0%	791,460.8	2.15%
Export	22,373.0	8,932.0	39.9%	968,935.6	2.31%

Source: UN (2008).

* HS 2002 SITC Code 87, vehicles other than railway or tramway rolling stock. This is the broader sectoral category, including vehicles and parts.

** Combination of HS 2002 SITC Sub-codes 8706–8708. Code 8706 includes chassis fitted with engines, Code 8707 includes bodies for motor vehicles, and Code 8708 includes parts and accessories of motor vehicles.

Table 6.5 Value of China's trade in vehicles and vehicle parts with South Korea and Japan, 2006 ($US millions)

Partner/Commodity	*Vehicle sector trade (share of total China vehicle trade) (A)*	*Parts sub-sector trade (B)*	*Vehicle/Parts (A/B)*
Import from South Korea	2,115.1 (12.4%)	1,539.5	72.79%
Export to South Korea	483.1 (2.6%)	315.7	65.35%
Import from Japan	5,309.2 (31.14%)	3,421.2	64.44%
Export to Japan	2,055.9 (9.19%)	1,261.8	61.38%

Source: UN (2008).

* HS 2002 SITC Code 87, vehicles other than railway or tramway rolling stock. This is the broader sectoral category, including vehicles and parts.

** Combination of HS 2002 SITC Sub-codes 8706–8708. Code 8706 includes chassis fitted with engines, Code 8707 includes bodies for motor vehicles, and Code 8708 includes parts and accessories of motor vehicles.

Toyota and Denso

Discontinuities in corporate organization likewise have proved difficult. Toyota developed facilities in Tianjin initially through a Daihatsu tie-up with the city-owned Tianjin Automobile Industrial Group. Toyota usually would first establish parts supply networks before investing abroad in engine plants or assembly, but the case of China was different. One condition for Toyota's joint-venture with Tianjin Automobile was supply through the firm's internal network located within the city and environs. Denso officials at the time complained about the difficulty of upgrading production among an existing set of local suppliers (INT 1999c; 1999f; 1999g). Efforts to promote a model of "cooperative learning" abroad usually begin with a "cooperative (*kyōryokukai*)" of first- and second-tier suppliers. Denso leaders in Tianjin through the 1990s had difficulty in identifying eligible local firms, let alone in cooperating with them in developing specifications and production processes.

Toyota entered the China market only in 1997, well behind their Western competitors, and then only in Tianjin rather than Shanghai to the south with its extensive array of assemblers and supporting industries. But today, Toyota appears to be faring well in the competition. Toyota produced 2,000 units in China in 2001, and sold 12,000. In June 2002, FAW acquired Tianjin Automotive Company, Toyota's original joint-venture partner, and helped to change Toyota's fortunes (INT 2002c). Five years later, the firm produced 280,000 units and sold 317,700 (Fourin 2007: 96). Toyota ranked sixth among leaders in the domestic market by 2007 with a nearly a 7 percent share (Fourin 2008: 213). Their flagship joint-venture with FAW operates two plants with 12,000 employees in Tianjin with annual production capacity of 200,000 vehicles (JAPIT 2007: 332).

Toyota China includes fifteen other major investments in auto manufacture, almost all joint-ventures with at best a 50 percent share. A Toyota family of firms on the ground in China supports the mother firm, including Hino Motors for trucks, Daihatsu Motors, and suppliers such as Aisin, Toyoda Gosei, Denso and others. Denso alone operates with twenty-four joint-ventures, five of which employ 700 or more workers. Flagship firms operating from the mid-1990s include Tianjin Asmo, Tianjin Denso Electronics, Tianjin Denso Air Conditioner, and Guangzhou Denso (JAPIT 2007: 328). Toyota has made a huge commitment to the China market, beginning with the parts industry in the 1990s, and building up to assembly late in the same decade, followed by alliance with the powerful FAW in 2002 and rapid expansion of production. FAW and the local auto industry have anchored Toyota production. Nikkei firms operated 100 research institutes in China at the end of 2007, mainly for product development. The tendency of Japanese firms to locate R&D facilities close to manufacturing or sales bases is particularly prominent in China (MERI 2008). As well, Toyota has begun to anchor product development close to its production plants, and training close to dealerships.

Complementarity

Deep historical ties with Korea, including colonization, and then a major Japanese role in Park Chunghee's development plans after 1965, have fostered complementary structures of Asian business in Japan and South Korea (McNamara 1990; 1999). Sustained efforts by successive Japanese governments have brought an overlay of trade associations, SME support structures, and FDI policies conducive to supporting investing Japanese firms in Thailand. But neither the Japanese state nor Japanese firms have made much progress in building complementarities in corporate governance with partners in China, in exchanges with trade associations, or in the legal framework of investment.

Nikkei investors continue to complain of weak intellectual property rights (IPR) protections, inconsistent enforcement of contracts, and confusing regulations. The situation appears to offer little prospect for Nonaka and colleagues' context that "harbors meaning," providing a common time and space for sharing knowledge (Nonaka, *et al.* 2001b). Yet defying expectation, we find a growing number of Nikkei research institutions for knowledge exchange, and growing competition with auto firms from the West and also from South Korea to tap local knowledge workers.

Structural discontinuities between state and private sector in China and Japan were apparent from the outset of trade ties in the early 1970s. Government officials organized the Japan–China Economic Association from 1972, with the Ministry of International Trade and Industry (MITI) playing a leadership role in a trade association for the private sector. Then again in 1994, MITI initiated a set of bilateral conferences under the Japan–China Automotive Industry Development Council. China's industry leaders welcomed the opportunity to learn everything about Japan's driving infrastructure, including insurance, driver education, and other supporting mechanisms. Japan's auto industry leaders hoped to play a major role in infrastructure development, to gain entrance to the market, and to develop a network of parts suppliers. The exchange ended when it became apparent that Japanese firms would continue to face restrictions on moving production facilities to the peninsula (INT 1999g; Nichū 1995).

China's state administration played a major role in the conferences. Industry and trade associations such as the China Council for Promotion of International Trade (CCPIT) continue to work today in concert with state officials, but have maintained four decades of ties with major Japanese associations. Together with Japan's leading trade association, the Keidanren, the Japan–China Economic Association (JCEA) under the direction of METI in Tokyo remains the most prominent of Japan's associations on the continent. Few Japanese firms can challenge Toyota's prominence in industrial diplomacy. Hiroshi Okuda served as Toyota chairman from 1999 to 2005, and chairman of the Keidanren from 2002 to 2006. Fujio Cho succeeded Okuda at Toyota, and served as vice-president of the Keidanren from May 2005, and of JAMA from 2006. Cho assumed leadership of the JCEA in March 2007 (Xinhua 2007). The prominence of the JCEA in bilateral economic relations was evident in the meeting of Chairman Cho with China's President Hu Jintao in August 2008 (CCPIT 2008).

for nearly 8 percent of the total value of commodity exports (Table 6.6). But better than 75 percent of the $10 billion in exports were vehicles of foreign auto assemblers from a Thai base. Vehicle sector imports totaled nearly $4 billion, 3 percent of total commodity imports. Parts accounted for 72 percent of total imports in the sector, amounting to $2.8 billion. Just less than $2 billion worth of parts were brought in from Japan (Table 6.7).

A major trade deficit in the sector draws attention to Thai dependence on the Japanese industry. Vehicle and parts exports to Japan of $500 million appear meager when imports top $2.5 billion (Table 6.7). Thai firms do find a role in manufacture of parts. Parts accounted for one-quarter of total Thai vehicle and parts exports in 2006, indication of a small but growing role for Thai parts makers (Tables 6.6). But Japan imported only 15 percent of the $2.5 billion in total Thai parts exports in 2006, destined rather for lower-technology auto sectors such

Table 6.6 Value of Thailand's trade in vehicles and vehicle parts, 2006 ($US millions)

Partner/Commodity	*Vehicle sector trade (A)**	*Parts sub-sector trade (B) ***	*Parts/ Vehicle (A/B)*	*Total Thai commodity trade (C)*	*Vehicle/ Total (A/C)*
Imports	3,916,834.5	2,849.4	72.7%	128,584.4	3.05%
Export	10,093.0	2,511.8	24.9%	130,580.0	7.73%

Source: UN (2008).

* HS 2002 SITC Code 87, vehicles other than railway or tramway rolling stock. This is the broader sectoral category, including vehicles and parts.

** Combination of HS 2002 SITC Sub-codes 8706–8708. Code 8706 includes chassis fitted with engines, Code 8707 includes bodies for motor vehicles, and Code 8708 includes parts and accessories of motor vehicles.

Table 6.7 Value of Thailand's trade in vehicles and vehicle parts with China and Japan, 2006 ($US millions)

Partner/Commodity	*Vehicle sector trade (share of total Thai vehicle trade) (A)*	*Parts sub-sector trade (B)*	*Vehicle/Parts (A/B)*
Import from China	123.1 (3.14%)	42.1	34.3%
Export to China	66.8 (0.66%)	50.4	75.4%
Import from Japan	2,539.0	1,994.7	78.6%
Export to Japan	505.0	372.2	73.7%

Source: UN (2008).

* HS 2002 SITC Code 87, vehicles other than railway or tramway rolling stock. This is the broader sectoral category, including vehicles and parts.

** Combination of HS 2002 SITC Sub-codes 8706–8708. Code 8706 includes chassis fitted with engines, Code 8707 includes bodies for motor vehicles, and Code 8708 includes parts and accessories of motor vehicles.

as Malaysia, Indonesia, and Australia (Oizumi 2008b). Exports to China included mainly parts, but as yet in relatively small amounts. In contrast, imports of Japanese parts amounted to nearly $2 billion, 78 percent of total vehicle and parts brought in from Japan (Table 6.7).

Toyota and Denso

Toyota Motor Thailand (TMT) began production at Samut Prakan outside of Bangkok in December 1964, and employs nearly 12,000 workers today. Toyota Auto Body followed from 1979, then Thai Auto Works with 500 workers, and Siam Toyota Manufacturing with 2,100 workers in 1989. More recently, Toyota established Toyota Technical Center Asia Pacific at Samut Prakan in April 2005 with a staff of 300. Functions include product design, testing and evaluation, with the goal of "modifying the specifications and body to reflect the tastes of each market." Then in July 2007, the firm established Toyota Motor Asia Pacific Company in Bangkok to provide production, purchasing, and logistics support for operations across the Southeast Asia region. Far and away the dominant auto firm in Thailand, Toyota sold 300,000 vehicles in the domestic market in 2006, representing a market share of 43 percent (Ward's 2007: 24, 59). The mother firm retains ownership, apart from a small share for Siam Cement in Siam Toyota, and for the Bangkok Bank and others in the huge TMT (TKS 2007a).

Toyota plays such a central role that that its own progression offers a chronicle of the local industry in initial production in the 1960s, expansion in the late 1980s, full ownership and reinvestment in the Financial Crisis, and then the shift to export markets in the twenty-first century. Deep roots across five decades are now reflected in the Toyota family firms operating in Thailand as well. Among its major suppliers in Thailand, all are Nikkei firms, and many Toyota firms, including Siam Toyota, Denso, Toyoda Machine Works, Siam Aisin, and Toyoda Gosei (IRI 2005).

For instance, Denso (Thailand) was established in 1972 for spark plugs, and has evolved into a firm with 3,000 workers and a range of electrical automotive components. Fifteen years later a Denso investment in local production of dies and jigs marked a new stage of localization for the Toyota family networks. With the shift to exports and higher technology, Denso opened wholly owned subsidiaries in 2002 for fuel injection systems and oil filters, including Siam Denso with 2,500 workers (Denso 2008c). Denso expanded with a new technical center in 2007 to supervise local production and provide training, as well as to develop products (Denso 2007). Thai Denso listed six manufacturing companies, and two business administrative firms with a total of 7,700 employees in Thailand in 2007 (Denso-T 2008). Toyota and Denso worked in tandem to develop Thailand as a regional hub for assembly, and now for export of one-ton pick-up trucks and multi-purpose vehicles under the International Multi-purpose Vehicle (IMV) Project. Serving as one of Toyota's bases, the project has propelled Thailand into the worldwide race among major automotive export platforms (Osono 2006b; Osono, *et al.* 2008).

Beyond the firms themselves, business–state coalitions have been central in the formation of trade policy, and in aid-supported training programs for upgrading production. Both firms promoted the "Brand-to-Brand Complementation Scheme" in ASEAN nations, which reduced tariff barriers for the flow of parts within the region (INT 2002b). The ASEAN Free Trade Agreement (AFTA) then permitted more extensive reductions. Export products with the highest utilization rate of AFTA benefits include automobiles, pick-up trucks, air-conditioners and refrigerators. Thai exports in 2007 under the AFTA tariff-reduction scheme amounted to $7.8 billion alone, with total exports to ASEAN nations jumping 28 percent over the previous year (BP 2008a). Additional trade agreements have opened trade channels beyond the ASEAN nations. Toyota and Denso have now begun to reorient to markets in India following ratification of the Thai–India FTA. Regional trade contributes to the anchoring of Nikkei auto manufacturing in Thailand. Far more significant is the Japan–Thailand Economic Partnership Agreement (JTEPA) itself, an opportunity to radically deepen Nikkei networks for auto production in Thailand.

Both firms have also long promoted training programs in Thailand, usually funded by the Japanese government. Toyota Motor Thailand CEO Mitsuhiro Sonoda plays the traditional dual role of head of TMT and of the Thai–Japanese Chamber of Commerce. He emphasized the need to improve human resources in Thailand, and the benefits of JTEPA for lowering tariffs on vehicles and parts imported from Japan. Noting however that Toyota has achieved 90 percent parts localization, he pointed out that the firm imports only "the automatic gearbox, 4-x-4 transfer case and steel (TN 2007a)." He concluded the interview with a plea for advanced engineers in the fields of materials, electronic and mechanical engineering. Founded in 1954, the Japan Chamber of Commerce and Industry (JCCI) in Bangkok lists some 1,500 members, including seventy-four transportation firms, and remains the largest Chamber outside of Tokyo and Osaka.

Complementarity

Japan has worked long and hard to establish complementarity with social production systems in Thailand, as is evident in aid programs from 1967. Aid amounts totaling over $9 billion were cited in early JTEPA negotiations (MOFA 2003). A Japan International Cooperation report of 2003 made clear Japan's prominence in Thai foreign relations: "Japan ranks first among Thailand's trade partners, foreign investors, and bilateral donors (JICA 2003a: 19)." Support included economic assistance consisting of grant aid, technical cooperation, and yen loans for infrastructure projects. With the policy shift in 2003 from "development assistance" to "new cooperation," Japan hoped to strengthen links with the growing private sector.

Industry associations can play a major role in facilitating such links through training programs, standardization efforts, electronic commerce, and negotiations with other industries (Doner, *et al.* 2004). Thais have not been able to emulate Japan's model of business–state cooperation through cohesive industry associations, being unable to provide effective partners for the Keidanren, JCCI, or JAMA.

Nonetheless, the Japanese state and industry have worked closely with Thai officials and business leaders in law, industrial policy, and training (INT 2002a).

Thailand joined the Trade Related Aspects of Intellectual Property Rights (TRIPs) Agreement under the WTO, though largely due to pressure from trade partners rather than to a domestic consensus (Sukonthapan 2005). Efforts to promote Thai commitments to property protections have succeeded to the point that Japanese firms no longer list information leakage among their major concerns (JBIC 2007: 18). JTEPA includes "technical cooperation activities related to the implementation of competition law and policy (MFA 2007: Art. 15)."

Japanese programs to support financing and training at Thai SMEs have been a long-standing priority in projects to influence local industrial policy. A METI Report of 1999 under Shiro Mizutani outlined Japanese priorities for infrastructural support of SMEs in Thailand (TN 1999). Four years later in JTEPA negotiations, METI officials cited achievements such as the "Automotive Technology Building Program," the establishment of the Office of SMEs Promotion, the enactment of the SMEs Promotion Act, and the SME Management Consultant (SHINDAN) System (MOFA 2003). It is thus no surprise that SMEs take center stage in Chapter 11 of the JTEPA Implementation Agreement, but with a new emphasis. Moving beyond assistance for "capacity" issues, Article 50 promotes new linkages between Japanese and Thai SMEs beyond Thailand, "to cooperate in their businesses particularly in Southeast Asia." The language gives a legal form to the evolution of Nikkei SMEs in Thailand from local to regional suppliers in tandem with networks of Thai sub-contracting partners.

METI officials cite five decades of technology transfer through dispatch of experts to Thai factories in a sustained effort to develop supporting industries. State-supported programs such as the Association for Overseas Technical Scholarship (AOTS) sent Japanese engineers to Thailand already from the 1960s. Then in 1972, Japan established the Japan–Thailand Economic Cooperation Society (JTECS) in Tokyo, and the Technology Promotion Association (TPA) in Bangkok the following year (TNI 2007). Among the most recent efforts, Toyota and the JCCI Bangkok established the Thai-Nichi Institute of Technology (TNI) in Bangkok, including a small four-year college program with Engineering, Information Sciences, Economics and Management departments. Improvement of Thailand's supporting industries appears to be the basic goal of the Institute (JAMA 2007b; 2007c; JCCI-B 2007b).

Within the auto industry, Japanese state and private enterprise have joined in efforts at standardization and human resource development through the Thai Automotive Institute (TAI). Thai officials in tandem with Japanese advisors founded TAI in 1998 as an organization of "knowledge and expertise" for the development of the auto industry. TAI operates today with a staff of 150 technicians for product testing in parts manufacture, establishing standards, and providing training in production process and organization management (TAI 2005; 2008).

JTEPA heralds a new era of complementarity in the auto industry. In brief, Japan agreed to open up markets for textile and apparel, and Thais in turn agreed to ease restrictions on autos and auto parts, and iron and steel from Japan. Partners agreed to implement seven "cooperation projects," including the Automotive Human Resources Development Institute in tandem with the TAI. A consensus was reached on the goal of "turning Thailand into a world class sustainable production base in the automotive sector, and solve the problem of skilled labor constraints through human resources development (MOFA 2005)." Negotiators enumerated a progression from outfitting an Institute with advanced equipment and a training curriculum, then dispatching Japanese automotive experts, and finally integrating the program with universities and vocational colleges. Japanese offices included METI, JETRO, and JAMA (Jirapet and Amari 2007a).

Prospect for knowledge networks

A place of production and a space of knowledge exchange appear critical factors in the emergence of knowledge networks. For instance, placement of facilities affects the space of exchange. An absence of Japanese assemblers in place producing on the Korean peninsula limits exchange in the space of cross-border knowledge flows. On the other hand, space can affect placement of production. Toyota's relatively late arrival to the manufacturing competition in China was due in part to fears about weak intellectual property protection on the continent (INT 1999f; 1999g). Thailand appears well positioned to capture a place in the space of Nikkei knowledge flows, given the extensive manufacturing investment of Japanese auto firms. Place appears to foster a space of knowledge flows. Yet weak complementarity between social systems of production in Japan and Thailand impedes more extensive exchange. Scholars point to the weakness of National Innovation System in Thailand, ineffective business associations, and inadequate support for SMEs. Complementarities of place deeply affect the space of knowledge flows.

This is not to deny the remarkable evolution of the Thai auto sector from import substitution to a dynamic export sector in the past decade, or the growing sophistication of auto production. Executive Vice-President Thanatorn of Thai Summit, a major local parts firm, sounded confident that with additional R&D, local manufacturers could meet specifications for eco-car projects (TN 2007f). Certainly there is potential, but production and trade data do not necessarily reflect the more embedded process of deepening knowledge exchange across national borders. We cited other indicators above, such as patenting activity or national budgets for research, with Thailand significantly behind its neighbors to the north. What then is the prospect for knowledge exchange in Thailand?

The extensive Nikkei investment already in place, as well as access through the Thai gateway to markets in Southeast Asia and India, and the security of Thailand as a "China + 1" all hedge against uncertainties on the continent and contribute to the anchoring of Nikkei manufacturing in Thailand for the foreseeable future. The question remains whether the nation can anchor Nikkei innovation.

Two recent events may well determine the direction of knowledge exchange between the two nations in the auto industry, one at the national level, the other at the sectoral level. Conclusion of JTEPA promises to bring additional funding and Nikkei expertise to Thai SMEs, the core of the nation's "supporting industries."

More extensive partnering with Japanese SMEs in Thailand will encourage further complementarity and cooperation. Cooperative learning between larger and smaller firms in Japan continues to provide highways for knowledge exchange. Technologically sophisticated Nikkei SMEs may be well positioned to offer similar opportunities to local SMEs with learning capabilities. However, not all share such optimism. Secretary-General Thavorn of the Thai Auto-parts Manufacturers Association (TAPMA) went public with his fears that JTEPA would reduce technology transfers from Japan, and indeed might replace local production with imports from Japan (BP 2007a).

At the sectoral level, Toyota's IMV Project may provide a new format for linking local minds and global modules in auto production (Osono, *et al.* 2008). The initial Thai role appeared limited to design of the IMV at Toyota's research center in Samut Prakan. Toyota engineers dispatched from headquarters largely designed and set the manufacturing specifications for the pick-up truck module. If the project is successful, however, Thai engineers will probably gain a greater role in this global module effort as localization proceeds. Toyota investments in engineering education programs of the Thai-Nichi Institute, and hopefully extension of auto engineering programs to universities and vocational schools, would provide a stream of trained personnel for production of the modules for export as well. Place makes a difference, and in this case suggests potential for knowledge exchange, assuming complementarities can be improved.

Conclusion

Innovation in integrated production demands far more extensive continuities in social systems of production than we found in the modular production of electronics. South Korea has evolved with the deepest structural continuities with Japan, whether in corporate structure, national innovation systems, industry associations, or in sector-specific technology policy. China presents a challenge for conservative foreign investors such as Toyota, given extensive discontinuities in corporate structure and ownership, state controls in industry associations, and local priorities of localization and indigenous access to foreign technology (Cohen 2007; Gao 2000; INT 2007d). Thailand's long-standing ties with Japan present a quite different profile and prospect. Japan has long promoted continuities in financing and training for SMEs, in joint programs with industry associations and the TAI, and now with JTEPA in corporate organization, training, and technology transfer. Disparities in national innovation systems, support for R&D, and the supply of engineers provide a sobering counterpoint to the promise of technology flows.

Social systems of production

Regional integration has promoted some standardization of structure and practice in both electronics and auto sectors, although far short of JETRO hopes for a "seamless market" across the region. Interviews with Japanese representatives of the Asian Business Advisory Group under APEC highlighted inefficiencies due to discrepant national standards on auto parts and electronics products (INT 2007b; 2007c). Bi-national committees under JTEPA on SMEs and specific sectors will monitor efforts towards convergence of structure and practice between Japan and Thailand.

Convergence of standards between Japan and South Korea has already advanced due to common interest in protection of intellectual property, and on global market standards. Proponents of a Japanese trade agreement with South Korea look beyond convergence to the competitive benefits of a more efficient division of labor between the auto sectors in the two nations. South Korea has moved quickly to conclude trade pacts with ASEAN, the USA, and now India, although none has been ratified to date. Approval and implementation of these trade agreements may turn Korean innovation energies beyond Nikkei networks to other equally capable partners. Nonetheless, of the three target nations here, structural compatibilities and similar technology levels make South Korea an ideal partner for tapping mutual strengths in national innovation systems, and for knowledge exchange through alliances and affiliates.

Japan's challenge in China remains the wider transition from patron to partner in R&D, and the sectoral transition in autos from simple assembly to craft in integral manufacture. Localization of design and component manufacture would benefit both nations, despite persisting problems of incompatibilities in social systems of production. Unfortunately, political differences persist as fertile ground for techno-nationalism on both sides, particularly strong in major import-substitution industries closely identified with the nation. China's auto industry appears to be evolving towards global modules with only limited participation by local minds, due in part to IPR issues on the ground, and in part to the speed of model changes and technological advances among major foreign firms. Chinese partners have caught up with joint-manufacturing projects, but can they now make the step up to design and engineering of localized models? Beyond China, can local manufacturers affiliated with major Japanese makers "catch up" with a very competitive global industry? Place without strong bi-national compatibilities does not foster a common space for knowledge flows.

Global value chains

Automotive industries shed a powerful light on state and firm in the interaction of local system and global value chains. State strategies of import substitution for autos in the 1960s shifted quickly to export orientation for the industry barely a decade later. Rapid upgrading of production for export far outpaced industry formation in production of auto parts, forcing the state to rely on foreign producers for parts. At the same time, state localization strategies kept foreign automakers limited to smaller roles in joint-ventures. Local assemblers such as Hyundai–Kia have evolved into major global players, and Hyundai Mobis into a major first-tier supplier. Simultaneously, global events such as the Asian Financial Crisis forced the state to seek IFDI among failing assemblers such as Daewoo, Ssangyong, and Samsung. What results today is an industry closely linked to global networks through both strong local firms and leading Western global assemblers. South Korea's auto industry enjoys strong technology links to global knowledge flows in the industry.

State-directed trade policy promises yet more changes, with the prospect of FTAs with ASEAN, India, and the USA. Trade pacts would permit a deeper flow of technology exchange. At the corporate level, foreign-invested firms such as Renault–Samsung, or GM Daewoo on the peninsula already join knowledge flows among Japan, South Korea, and the West. Hyundai–Kia would be the exception without transnational capital links to major global firms, but with strong prospects for alliances to establish new flows. Ironically, this profile has developed with Japan largely absent, whether in trade agreements, or at the corporate level, apart from Renault's tie-up with Nissan. Although linked in the auto parts sub-sector, weak links with Toyota and Nissan on the Korean peninsula impede broader knowledge flows among the two nations with the strongest continuities in social systems of production among the four nations of our study.

China opted for direct foreign investment to jump-start their import substitution in autos, in contrast to the Korean state strategy favoring IDFI through technological agreements. Only today do we find the beginnings of a shift from import-substitution industry towards an export-oriented industry, in part due to the attractiveness of the huge domestic market. Cultivating a foreign-invested industry permitted a rapid development of auto production, but at the price of extensive reliance on foreign technology across the whole range of craft in integral production. South Korea has been able to develop craft to the point of reducing reliance to only high-technology parts and components. Chinese partners in foreign-invested firms have much to learn before their local industry can gain the independence of the Korean industry. Yet reliance may permit knowledge flows, if the Chinese state can play a constructive role in encouraging localization, as well as providing a national innovation system with the necessary human resources for learning.

Thailand has carved out a third path of state and corporate strategy *vis-à-vis* global chains in auto production. Import-substitution strategies brought Japanese producers in joint-ventures, and permitted development of supporting industries. Financial constraints forced most of the local partners out of the ventures in the late 1990s, but preserved the now foreign-invested industry. A subsequent shift to export orientation among the firms was supported by state policy, and has permitted expansion of the industry. But Thailand's further advantage has been aggressive trade liberalization, first within ASEAN, and now within the region. Neither China nor South Korea has emulated Thailand's close ties to Japan, now formalized in JTEPA. Structural discontinuities persist, such as a lack of competitive local assemblers, cohesive business associations, or a vibrant infrastructure of SMEs, but deep links to Japan offer potential for knowledge flows in GVCs of benefit to both nations.

7 Textiles and fashion

Local minds and global designs

East Asia has gained fame as the world's textile factory. The region operates much of the global capacity in synthetics and natural fibers, and churns out huge volumes of garments as well. China alone accounted for nearly 30 percent of the total value of global textile production in 2006, Japan 5 percent, Thailand and South Korea each about 2.5 percent. China's apparel sector generated one-quarter of all value produced in garments worldwide, Japan 10.5 percent, Thailand 6 percent, and South Korea 1.5 percent (UNIDO 2008: 55).

Textiles helped to pioneer Nikkei industries at home and abroad. For instance, Sakichi Toyoda gained fame with his invention of an automatic spinning loom, the centerpiece of Toyoda Automatic Loom Works founded in 1918. His son Kiichiro Toyoda spun off the combustion engine department of the firm as Toyota Motor Company in 1937. Early textile firms such as Toyobo and Toray contributed to industrialization across the region with plants in China and the colony of Korea in the 1920s (McNamara 1995). Shanghai prospered with a multinational mix of cotton mills early in the twentieth century, and then served the nation as a leading industrial sector under Communist leadership from 1948. Japanese textile plants reverted to local ownership in South Korea from 1948, the origins of a leading sector in the nation's export-oriented industrialization (EOI) a decade later (McNamara 2002). Major Japanese trade and textile firms moved to Thailand in the 1960s, playing a central role in founding the local industry.

Craft and technical expertise flow across an industry of high technology in chemical fibers (i.e., nylon, polyester, and acrylic) upstream, mid-range technology in spinning and weaving, and labor-intensive cutting and sewing downstream. Unlike a trading firm, a manufacturer builds knowledge networks around its own product expertise. Toray Industries is the nation's largest manufacturer of synthetic fibers. Textiles account for nearly 40 percent of sales, with chemicals, plastics, and carbon fibers providing a further 20 percent. Toray employs 55 percent of its workforce abroad, with more than 21,000 at work in 126 subsidiaries. Sales abroad accounted for 45 percent of total sales in 2007. The firm invested $458 million in research that year, amounting to 2.7 percent of annual sales (Toray 2008a: 34).

Japan's general trading companies (GTC), *sōgō shōsha*, carried yarn and fabric abroad already in the late nineteenth century, and returned with cotton. Traders today integrate the textile stream, purchasing fiber or fabric, sub-contracting for processing into garments, and then marketing as fashion under leading brand names. Traders also move with Japanese makers abroad in joint-ventures for local manufacturing, providing financing and marketing expertise. Third, GTCs can invest with local partners abroad as "venture merchants," whether through extensive sub-contracting networks, or in joint-ventures with major local manufacturers (Todeva 2006: 177). GTCs bring technology, technical expertise, and craft abroad, a major local benefit in their knowledge networks. And textiles continue to generate revenue. Trade in the sector generated 8 percent of annual sales at Itochu in 2007 and a net income of $144 million. Itochu listed total assets in the sector of $3.4 billion (Itochu 2007c: 28).

Necessities of food, shelter, and clothing command attention in the early phases of economic development, well before electronics or automobiles. A textile industry often plays a major role in initial industrialization, first for import substitution, and then for export-oriented strategies. All four nations in this study have followed that path, and all four rank today among the world's major textile exporters. They have all advanced into more modular production upstream of fibers and fabric, and then more integral production strategies in garments and fashion, where design, cutting, and assembly demand craft. Prominent Nikkei networks of trade and textile manufacture in South Korea, China, and Thailand have developed with extensive flows of technology and craft, but with different levels of local participation. All three nations have anchored Nikkei production in textiles, but can they anchor local minds in flows of global designs?

South Korea: from cooperation to future coordination

State and private sector cooperated in the 1960s on a plan to advance upstream from natural fibers to man-made fibers. State investment in heavy and chemical industries early in the next decade extended the project. Investments in polyester production proved opportune for a textile industry moving from cotton to man-made fibers. Toray Saehan, for instance, can procure the basic chemical for polyester production, terephthalic acid (TPA), entirely from local sources (KOTRA 2003).

The original plan succeeded beyond expectations, resulting in an integrated line of production from chemical fibers through spinning and weaving to garments and fashion. Emergence of South Korea's general trading companies in the same decade boosted textile exports. Organization of the GTCs, precedents of industrial organization from the colonial period, and reliance on Japanese synthetics firms for technology and expertise from the 1960s permitted continuity between textile sectors in the two nations. Direct investment by Japanese firms in production on the peninsula would strengthen those continuities in the social systems of production. Complementarity in the sector has supported extensive cooperation, and may lead to closer coordination in joint projects with a preferential trade agreement.

Automobiles and electronics came to overshadow the textiles and apparel within South Korean industry, just as competition from lower-wage economies abroad eroded Korea's cost advantage. South Korea listed 14,000 firms in textiles and apparel in 2005 with total employment of 185,000 workers. An additional 100 firms in man-made fibers employed only 6,500 workers in more technology-intensive production. The value of total output in textiles and man-made fibers ranked behind only electronics and transport equipment (UNIDO 2008: 583).

South Korea's spinning mills reported 3.3 million spindles in operation in 1985, but just better than one million spindles on the peninsula in 2007 (SWAK 2008). Among the first sectors forced abroad, spinners moved quickly to Indonesia and Vietnam and later to China. Smaller and more mobile, SMEs in the apparel sector moved as well. Synthetics firms continued to prosper until the past decade when competition with China from below and Taiwan and Japan from above forced restructuring. Installed capacity for manufacture of man-made fibers fell from 3 billion tons annually in 2001 to only 1,500 tons in 2007.

Despite consolidation, South Korea still ranks as the world's sixth largest producer of synthetics (JCFA 2007; JCHEM 2008: 447). The vast majority of textile and chemical operations remain local enterprises, including major firms such as Hyosung, Huvis, and Taekwang Industrial. Yet the profile of Japanese investment in South Korean textiles differs dramatically from autos and electronics: Nikkei firms find a place among Korea's leading mills. Among the ten members of the Korea Chemical Fibers Association (KCFA), Toray operates Toray Saehan as a subsidiary, and partners with Mitsui in Kolon Fashion. Teijin has a technical agreement with Seong-an Industrial Synthetics. Chonbang Gunze and Dongil Renown operate as major Nikkei firms in midstream and downstream segments (TKS 2007b). Cooperation in textiles extends to investment and production, and holds promise for coordinating a new regional division of labor.

Textiles and fashion

South Korea ranked fourth worldwide in textile exports in 2006 after the EU, China and the USA, with nearly a 5 percent share (WTO 2007). Yet auto and electronics exports far outpace a textile sector where $12 billion worth of exports still accounts for less than 4 percent of South Korea's commodity exports (Table 7.1). Synthetics filament and fiber exports of $4 billion, and fabric exports of nearly $5.5 billion indicate a dynamic, competitive sector despite intense competition in the region. Exports in that year of polyester staple alone earned $710 million (KCFA 2008: 60). China has eroded but not eliminated the South Korean fashion industry. Apparel exports suggest successful penetration of fashion and specialty markets where the Korean makers can establish a niche above the lower-cost garment producers in the region (KOFOTI 2008). On the other side of the register, apparel from abroad constitutes more than half the value of textile imports, although the sector still registers a positive trade balance.

If Japan has invested in South Korean textiles, South Korea has not returned the favor. Trade follows investment to some extent. Hyosung operates four spandex plants in China, and three other plants for tirecord. Taekwang, Kolon, Hyosung, Huvis, SK Chemical and Honam Petrochemical all operate plants in China as well (INT 2003). A Korean textile network has emerged across the region to match

Table 7.1 Value of South Korea's trade in textiles and apparel, 2006 ($US millions)

Commodity/Trade	*Imports*	*Exports*
Synthetics (A)	1,469.8	4,079.9
Fabrics (B)	777.7	5,475.4
Apparel (C)	3,790.3	2,481.2
Total textiles (A+B+C)	6,037.8	12,036.6
Total commodity trade	309,379.4	325,457.2
Textiles/Total trade	1.95%	3.70%

Source: UN (2008).

* HS 2002 SITC Code 54–63. I include in the category of synthetics man-made filaments (Code 54) and man-made fibers (Code 55). In the category of fabrics, I include wadding (56), carpets (57), wovens (58), coated fabrics (59), and knits (60). In the category of apparel, I include knitted apparel (61), general apparel (62), and other made-up textile articles (63). My category of textiles includes amounts from all ten codes.

Table 7.2 Value of South Korea's trade in textiles and apparel with China and Japan, 2006 ($US millions)

Commodity/Trade	*Import from China*	*Export to China*	*Import from Japan*	*Export to Japan*
Synthetics (A)	590.1	921.2	172.8	89.7
Fabrics (B)	247.5	1,223.4	128.8	199.4
Apparel (C)	3,006.1	402.5	62.4	335.2
Total textiles (A+B+C)	3,843.7	2,547.2	364.2	624.4
Total commodity trade with partner	44,522.2	89,724.1	50,270.0	27,328.4
Textiles/Total trade	8.6%	2.8%	0.7%	2.2%

Source: UN (2008).

* HS 2002 SITC Code 54–63. I include in the category of synthetics man-made filaments (Code 54) and man-made fibers (Code 55). In the category of fabrics, I include wadding (56), carpets (57), wovens (58), coated fabrics (59), and knits (60). In the category of apparel, I include knitted apparel (61), general apparel (62), and other made-up textile articles (63). My category of textiles includes amounts from all ten codes.

product to cost. The shift began with garments as mobile SMEs moved into China to take advantage of lower labor costs. Spinners and then man-made fiber firms moved to Indonesia and Vietnam, and finally joined chemical firms in the move to China.

China's geographical proximity to the peninsula permits low transportation costs for South Korean plants re-importing. Textiles, mainly apparel, accounted for nearly 9 percent of total South Korean imports from China in 2006 (Table 7.2). Filaments and fibers went back to China, but amounted to only about 3 percent of South Korean exports to its neighbor. Textiles played a far smaller role in trade flows with Japan. South Korea imports specialized fabrics and fibers, and exports mainly apparel. The fact that South Korea retains a market in Japan for apparel despite competition with lower cost producers in China suggests higher-value-added goods.

Sector and society

Japan's anchor in South Korean textiles begins with firms, extends to industry organizations, and may some day culminate in coordination of a new regional division of labor. The emergence of GTCs at leading South Korean *jaebeol* in the 1970s curtailed opportunities for Itochu, Marubeni and other textile-oriented Japanese GTCs in the Korean sector. More important was the role of makers such as Teijin and Toray in advancing the upstream, technology-intensive production of polyester, nylon and other chemical fibers. Toray has assigned higher-value-added manufacture to South Korean joint-ventures within an integrated system of production across East Asia.

Kikkawa contrasted global strategies in the 1990s of an electronics firm like Sony with a textile manufacturer like Toray (2000; INT 1999h). The former was content with "global products," but the latter opted for global operations, locating procurement and production close to markets. Toray's long-standing investments in South Korea, China, and Thailand appear to fit the thesis. Toray has long held a minority share in Kolon Industries, and joined Kolon as a minority partner to establish KTP Industries in 1998 for the production of resins (TKS 2007b: 132). But if Kolon is a textile and fashion partner, South Korea's largest *jaebeol*, Samsung, is partner in textiles and electronics.

Toray had long ties with Samsung's Cheil Wool, and its successor, Saehan Industries. Toray acquired Saehan's plant for polypropylene spunbond during the Asian Financial Crisis of 1997 with help from both the Japanese and South Korean governments. Spunbond has proved a market favorite for applications such as diapers and other sanitary materials. The venture has succeeded beyond expectations, with exports to Japan and US markets.

Toray Saehan boosted capacity in Gumi, South Korea, with an additional plant, and with a new subsidiary for spunbond and applications in Jiangsu Province, China, titled Toray Polytech (Nantong). Toray Saehan and Toray Inc. each hold 50 percent of the shares. Officials explained the rationale of a South Korean subsidiary producing in China: "[T]he Toray Group aims to take advantage of the technology, expertise and track record established at Toray Saehan in manufacturing and sales (Toray 2006)." Other investments in South Korea include a 50:50 joint-venture in precision instruments for synthetics production at a subsidiary called Daehan Precision, from 1988 (TKS 2007b: 132).

Chemical applications stretched beyond textiles as Toray joined Samsung's semi-conductor assembly expertise to their own strengths in base film. STEMCO

(Samsung Electromechanics) Ltd is a 70:30 venture to provide display applications for phones and other uses. Toray then invested as the minority partner in a 49:51 venture called STECO (Samsung Toray Electronics). STEMCO President Park Kyu-bok noted in 2004: "Toray supplies tape raw material to us, which we use to manufacture Tape Carrier Packaging, supplied to Samsung Toray Electronics Company to make TCP assemblies, which then goes to Samsung (KOTRA 2004)." The supply chain demands extensive cooperation in product development. As STECO President Park explained in 2006: "[A]fter building a direct connection with all related firms in terms of raw materials, designs, packages and chips, STECO dispatches core personnel to Japan's Toray, Samsung Electronics and film makers and closely cooperates with these companies in technology development (STECO 2006: 52)."

A society of industrial organizations helps to secure knowledge flows. Close historical ties between the sectors of the two nations have fostered complementary industry associations. South Korea's Spinners and Weavers Association (SWAK) meets annually with the Japan Spinners Association. Similarly the Korean Chemical Fibers Association (KCFA) consults regularly with the Japan Chemical Fibers Association (JCFA). On the labor side, predecessors of the Federation of Korean Textile and Distribution Workers Union met annually with Japan's UI Zensen textile union federation. As the South Korean textile industry moved abroad from the 1990s, the Korean federation reshaped itself as an international organization with support from UI Zensen (McNamara 1996; 2002). Japan's UI Zensen helped found, fund, and continues to foster the Textile Workers Asian Regional Organization (TWARO) within the International Textile, Garment, and Leather Workers' Federation. Today the Korean federation has shrunk to only 10,000 members, but continues to participate with Zensen in TWARO activities (ITGLWF 2008).

Prospect

Precedents of knowledge exchange suggest possibilities for future flows in Nikkei textile networks. Japan's Ministry of Trade and Industry (MITI) joined with leading firms and sectoral associations in "structural adjustment" from the 1980s (Yamazawa 1988; 1990). MITI managed programs with associations reducing capacity to avoid disruptive market fluctuation and "excessive competition," nudging firms out to other sectors, or up into higher-value-added textiles and fashion. State and firm mobilized regional associations in neighboring nations with a similar dual strategy of stabilizing markets and upgrading production through the exchange of information and even voluntary coordination of production.

MITI officials organized representatives from seventeen nations in the Asia–Pacific Textile Industry Forum in Kyoto in June 1996. Canada and the USA, China, Thailand, and South Korea, India and others joined to share data and discuss prospects for expansion. Annual meetings across the next five years tracked trade and investment and introduced new trends in technology (ASPAC 2002). Yoshimatsu cited loss of interest among member economies in continuing the dialogue as a factor it its demise, despite accomplishments such as an integrated database system

(2002a). Whatever the benefits of sharing information, exporting nations facing the imminent termination of the textile quota system in 2005 were probably less willing to coordinate across borders in the ensuing transition.

A region-wide project in synthetics has proved more durable. MITI supported Toray's CEO Katsunosuke Maeda, then President of the JCFA, in a project to mobilize counterpart associations of synthetics producers from nine nations in 1996 (Ishii 2005). What resulted was the Asian Chemical Fibers Industries Federation (Dempsey 1996). The nine original Federation members included China, India, Indonesia, Japan, South Korea, Pakistan, Chinese Taipei, Thailand and Malaysia. Survival or "co-existence," and co-prosperity served as initial goals.

The group finally hammered out their "Vision" only a decade later at a meeting sponsored by the KCFA (ACFIF 2006). Among priorities, the group committed itself to balance of supply and demand, development of consumer markets within their nations, and advances in technology and products. They agreed to address problems of commercial rights and intellectual properties. Finally, they proposed a free flow of personnel in the industry among member nations to improve the regional industry. Asian Chemical Fiber Industries Federation (ACFIF) members reaffirmed the goals and again debated issues of oversupply in the region at a subsequent meeting in Malaysia in May 2008 (BERNAMA 2008; KCFA 2006).

Man-made fiber plants demand huge investments which permit only larger firms in the sub-sector, unlike the wider textile and fabric industry with a large number of smaller and larger firms. Concentration contributes to better concentration and consensus within industry associations, and indeed within a regional group. An Asian Chemical Fibers Federation with a more select, smaller group of member firms could survive where the far larger, more diverse Textile Industry Forum could not. Larger firms in the Chemical Fibers Federation with better-funded national associations could also organize separately from MITI, relying rather on the JCFA as quasi-secretariat, and on other Associations chairing the bi-annual meetings. Cohesion, industry identity, and independence have supported their cooperation.

If the South Korean textile industry has played a prominent role in both regional associations in tandem with Japanese counterparts, stronger bilateral knowledge flows have been proposed in discussions of a Japan–Korea FTA. Kim Dohyung predicted a vertical intra-industry specialization in petrochemicals with increased sophistication of research in South Korea, improving the existing profile of Korea with ordinary products and Japan with specialized products (Kim Dohyung 2007). Extension of Japanese specialized chemical production to South Korea would spur competition and development, as well as joint exports to third countries.

But the sub-sector of fabrics and fashion might move towards even closer integration. South Korea has largely surrendered the clothing import market in Japan to China, a huge blow to the industry forcing specialization in the sub-sector towards sporting and leisure fashions. South Korea would need to upgrade fiber, fabric, and dyeing in Daegu, and fashion and logistics in Seoul at the Namdaemun Fashion area. Then joining with Japan's development of new fabrics and brand power, South Korea and Japan would jointly create an Apparel Valley at home, moving materials and sewing entirely to China. The two nations would collaborate not only in manufacture and distribution, but in global out-sourcing (Back 2005; Kim Dohyung 2007). Nikkei anchors in both sector and social systems of production for textiles and fashion on the peninsula may well permit deeper integration. Proposals in both sub-sectors move beyond cooperation to coordination of knowledge and production flows.

China: from consultation to cooperation

Limited ties beyond the nation's borders constrained demand to a domestic market through the 1960s. A long process of re-establishing ties began with Deng Xiao-ping's opening from 1972. Apparel soon offered an outlet for bringing the low-cost sector into export markets, in tandem with foreign investors. Japanese GTCs such as Itochu established vast networks of garment sub-contractors by the mid-1990s, and helped drive a new export-oriented strategy of industrialization.

State officials saw further opportunity upstream, rapidly expanding polyester production of fiber and blended fabrics to feed the growing apparel sector. Today the nation registers installed capacity for annual production of synthetics of 22,700 tons, 55 percent of world capacity. China more than doubled their nearest competitor worldwide with 10,000 tons of polyester filament exported in 2007, representing 27 percent of world export totals in filament (JCHEM 2008: 447). Milling of blended polyester-cotton fabric feeds a huge garment industry downstream that has redrawn the landscape of global trade in clothing. Nearly $300 billion of textiles and apparel amounted to 30 percent of the total value of China's exports in 2006, with $270 billion of apparel alone (Table 7.3).

Table 7.3 Japan share of China's trade in textiles and apparel, 2006 ($US millions)

Commodity/Trade	*Total imports*	*Total exports*	*Import from Japan*	*Export to Japan*
Synthetics (A)	6,622.0	12,131.1	1,552.5	219.7
Fabrics (B)	5,468.1	12,307.3	986.6	508.7
Apparel (C)	18,451.9	270,119.0	3,039.3	28,669.4
Total textiles (A+B+C)	30,542.1	294,557.5	5,578.5	29,397.8
Total commodity trade with partner	791,460.8	968,935.6	109,060.3	127,643.6
Textiles/Total trade	3.86%	30.40%	5%	23%

Source: UN (2008).

* HS 2002 SITC Code 54–63. I include in the category of synthetics man-made filaments (Code 54) and man-made fibers (Code 55). In the category of fabrics, I include wadding (56), carpets (57), wovens (58), coated fabrics (59), and knits (60). In the category of apparel, I include knitted apparel (61), general apparel (62), and other made-up textile articles (63). My category of textiles includes amounts from all ten codes.

China has emerged as a major node in the global apparel chain. The nation exported $220 million worth of synthetic fiber and filament, $500 million of fabrics, but a remarkable $28.6 billion worth of apparel to Japan in 2006. Apparel with some textiles amounted to 23 percent of China's annual exports to Japan, nearly one-quarter of total exports. Nikkei merchants like Itochu re-import and makers like Toray play major roles in the trade of cotton-polyester blended fabric and fashions, bringing back or "re-importing" apparel from Nikkei networks in China (INT 1999a; Ishii 2005: 6). Japanese firms have extensive roots in the sector. China has anchored Nikkei production to supply fiber, filament and fabric for the apparel industry. Moving upstream to fiber and fabric exports will demand more extensive flows of expertise.

Sector

China's commitment to a full-set industrial structure has paid handsome dividends in the textile and clothing trade. As the sector has forced adjustment in industries of neighboring countries, it has brought home employment, consumer items, and a trade surplus. Textiles and apparel ranked fourth in total output value in China in 2005, behind electrical machinery, industrial chemicals, and iron and steel. The sector ranked first in employment with 9.5 million workers, and first in number of enterprises with 34,500 registered firms (UNIDO 2008: 250). Domestic expansion of commodity polyester production has driven firms in neighboring countries to higher-value-added products upstream, as well as encouraging relocation of plants to China. Foreign textile majors from Japan such as Toray and Teijin, as well as competitors from South Korea and the West bring new technology and process to modernize the industry.

Unlike electronics and automobiles, a huge base in synthetics and especially in spinning and weaving predated the influx of foreign capital. Adjustment in the local industry proceeds in tandem with competition at home with foreign producers. Reforms in the sector deepened global knowledge flows. Chinese officials and firms have worked furiously across the past decade to expand capacity in man-made fibers, to reduce capacity midstream in spinning and weaving, and to upgrade and localize apparel downstream. Nikkei firms have played their role in fostering knowledge exchange, with production anchors in all three sub-sectors. Merchants such as Itochu downstream have brought expertise initially in moving fabric to garments, and now fabric to fashion with leading brands. Makers such as Toray brought sophisticated polyester production, and now non-wovens to China, together with technology and craft for market applications.

Nikkei merchants

Traders create value chains as they integrate the textile stream. Itochu has long been a leader among Japanese GTCs in China. Looking only to textiles, we find the firm engaged with processing and investment. A "processing trade (*kakō bōeki*)" begins with purchase of fiber or fabric, and continues with sub-contracting for processing into garments. GTCs then bring garments to market through large department stores and other retailers. Itochu began with a conventional bilateral stream in the 1990s, moving Japanese craft in design, cutting, and sewing to China, and then re-importing the garments to Japan (INT 1999i; 2004). The firm introduced Japanese technicians, new applications in finishing, and specialized products with a new flow of expertise across the two nations.

Hong Kong subsidiaries play major roles in procuring material and assembling apparel, including Itochu Textile Materials from 1998, and the venerable Prominent Apparel from 1983, the firm's oldest China venture. The GTC expanded dramatically on mainland China with three sole ventures, each today with 400–500 employees: Tianjin Huada in 1993, Qingdao Tri-Gents Clothing the next year, and Hangzhou Fujuitomi Silk Garments in 1995 (TKS 2007b). One key to successful processing is supervision, on-site inspection to insure quality, perhaps the most important contribution of Itochu to development of craft in the processing trade. Processing continues today, but in closer cooperation with leading local firms. Itochu recently formed a textile sales company in the USA with Shandong Weiqiao Textile Company, the largest cotton textile maker with 132,000 employees, and revenues of about $1.25 billion (Itochu 2008b; Weiqiao 2008).

The trading giant established Itochu Textile (China) in 2002 with similar scale and ownership, but far larger ambitions. Processing has evolved with what the firm terms a "symbiosis" of Japan's textile industry with China, allowing Itochu to "bring its integrated functions into full play (Itochu 2007a)." Moving beyond craft and technical expertise, the Shanghai firm has opened a new flow of ideas and images with global brands. Bridging manufacture and market, local and export sales, the Itochu Textile promotes leading brands at Ningbo subsidiaries near Shanghai, such as Le Coq Sportif and Renoma Apparel. What began as a re-import strategy of fabric and fashion for the Japanese market was now joined to a local market for Renoma Paris fashions.

Branding soon moved beyond the bilateral circle of Japanese fashion to Chinese manufacture to Japanese and local market. Headquarters has given marching orders to its Textile Division in China – parlay "existing brands (DESCENT, Le Coq), to deepen cooperation with powerful Chinese partners to set up new brands (Itochu 2008a)." A joint-venture in wool suits with the local Youngor Group in Ningbo from 2003 signaled ambitions to move up the value chain in global apparel exports. Two years later, Itochu allied with Youngor and Italy's Marzotto to provide high-end brands for the global market. Marzotto is a leader in fabrics, design, and brands such as Hugo Boss and Valentino (NFTECB 2005). A bilateral processing trade bringing Japanese expertise and fabric to Chinese assembly had evolved into a global processing trade of alliances among major global fabric and fashion firms in three nations (GMB 2006: xxix). A demand for higher-value-added fabrics may in time promote upgrading of local man-made fiber and midstream weaving facilities.

Nikkei makers

Trading firms take Japanese manufacture abroad. Itochu allied with Mitsubishi Chemical Corp. in a high-purity TPA project in Ningbo, the core material for polyester production (Itochu 2007b). The venture will operate with an annual production capacity of 600,000 tons. Unlike trading firms, textile firms such as Toray and Teijin in nearby Nantong have brought sophisticated upstream technology in polyester production to China. What makes Toray so special is the investment and expertise in bringing their high-quality fiber and filaments downstream into fabric for local and export markets.

The firm opened a sales office in Hong Kong in 1974, and then TAL Knits in 1990. TAL weaves yarn into fabric, dyes and then assembles garments for export to a third country, operating today with nearly 400 employees (TKS 2007b). Investment in mainland China itself began in 1994 with a joint-venture for weaving and dyeing at Nantong, a major textile district north of Shanghai across the Yangtze River. Investment progressed with a much larger subsidiary for polyester and nylon filament the next year, Toray Fibers (Nantong), in the same district. Equipped with annual capacity of 60,000 tons of polyester filament, the firm stands at the centerpiece of Toray operations in China (JAPIT 2007; KCFA 2008: 186; TKS 2007b).

Both firms have expanded across the ensuing decade, with the weaving plant employing 1,300, and the fiber plant nearly 400 workers. In addition to Toray Polytech for spunbond, Toray extended its capacity in midstream spinning, weaving and dyeing in 2006. Rather than a sole venture, the firm allied as lead investor with a major local spinner, JIFA of Qingdao. Toray holds 70 percent of the shares in Toray JIFA (Qingdao), with a workforce of nearly 900. An investment at JIFA in cotton-polyester textiles highlights Toray's focus on product development in China for local and export markets. The firm established Toray Fibers and Textiles Research Laboratories in Nantong in 2002, operating with a staff of nearly 200 including thirteen Japanese engineers and a capital of $16 million (JAPIT 2007; TKS 2007b). A focus on production process and product development at the labs indicates at least two kinds of expertise now moving across the Toray regional textile network in East Asia. Calibration of machinery, adjustment for different products, and problem-solving suggest craft. Product adaptation and new applications would demand more attention to design and development.

Complementarity

Weak structural continuities between sectors in China and Japan did not stem the flow of $29 billion worth of Chinese exports to Japan in 2006. Competition with South Korean and Western firms to move production to China trumped complementarity of association or corporate structure. China's huge textile sector has evolved with a mixture of state-owned and private firms, as opposed to private firms among Japan's textile moguls. China's textile associations have been formally separated from state bureaucracies only in the past decade, and remain tightly linked to state policy as semi-governmental groups. “Implementing state industrial policy” appears atop the list of activities for China’s Chemical Fiber Association (CCFA 2008; INT 1999j).

The Chinese Association has participated regularly in bi-annual meetings of ACFIF where neighbor country associations complain of over-supply due to China’s rapid expansion of polyester manufacture. Capacity for polyester staple fiber and filament grew from 2.1 million tons annually in 1996, to nearly 12 million tons by 2004, and then to 19 million tons in China by 2007 (KCFA 2008: 141). Consultation is a beginning. Actual cooperation in balancing supply and demand across the region, or perhaps exchange on technology advances would be the goal.

State and private sector in Japanese textiles continue efforts to forge complementarities at textile firms, associations, and state offices on the continent. Differences begin in corporate ownership. Diversity in the local sector, and adjustment, further complicate cooperative action with Nikkei networks. Larger firms in man-made fibers upstream monopolize much of the capacity in Japan and South Korea. Japan’s Chemical Fiber Association lists twenty members. China’s Chemical Fiber Association, on the other hand, registers 400 member firms, large, medium and small, private and public (CCFA 2008; KCFA 2008: 181–99). Adjustment under government-mandated programs to expand capacity upstream, and to reduce capacity midstream, must also be taken into account in planning for technology exchange or possible joint-investment. Weak complementarities may impede deeper knowledge flows among the organized interests of the sectors in the coming years.

Constraints on corporate activity among invested foreign enterprise have been eased over time. Nikkei textile firms have been permitted more access to the domestic market, though usually on the condition of expanding export activity. The trading company gained permission to establish Itochu (China) Holding Company in 1993. As noted in its brochure, “during the past dozens of years, Itochu has built a close cooperative relationship with the Chinese Government (Itochu 2008c).” China’s Ministry of Commerce approved the firm as a regional headquarters of a transnational company in 2005. Beyond Itochu’s own investing, financing, and leasing activities in China, the Holding Company can now provide the same set of services for firms outside of China. The firm operates from its Beijing offices with a capital of $100 million (Itochu 2008c; JAPIT 2007; TKS 2007b). Corporate structures make possible new linkages among knowledge networks, such as alliances among Italian and Chinese fashion houses.

Prospect

Only Teijin can rival Toray among Nikkei textile moguls at home and abroad. Neighboring plants in Nantong symbolize their odd blend of competition yet cooperation with downstream Nikkei SMEs. Both face growing competition from local Chinese firms. Teijin President Nagashima spoke recently of the Nikkei advantage in China: “[T]hanks to our technologies that have been accumulated over a long period of time, Teijin can add to products new functions that China has not yet developed (Nagashima 2007)." Both firms have staked their futures on parlaying advances in polymer chemistry into new consumer and industrial applications. But applications depend in part on local context. Indeed the "Common Tasks" of ACFIF include not only balancing supply and demand, but also cultivating "new markets," and "developing technology based on respective clothing culture (Ishii 2005: 13)." Nikkei firms will need local knowledge of market and consumer culture to succeed with new applications. Nagashima concluded that the biggest challenge for a global firm is "to secure human resources" on site abroad.

Local expertise extends beyond market and taste to industry. The Federation (ACFIF) operates with the dual goals of "coexistence for mutual benefit," and better corporate performance through information exchange (Ishii 2005: 13). How do information flows promote coexistence? Federation goals include "improving the efficiency of corporate activities by capitalizing on the macro information shared by individual companies." This presumes that individual firms within national associations are both capable and willing, and that national associations in turn can mobilize the firms to provide an accurate picture of the local industry and market. It further suggests common standards and categories, and a consensus on what information is necessary. Information exchange is a critical initial step for ACFIF, and a significant channel of knowledge exchange.

Preferential trade agreements promise to engage yet a third level of knowledge exchange that would link Nikkei networks in Southeast Asia to the China market. There is little immediate prospect of a bilateral agreement between China and Japan. But Thailand and China have already implemented the first stages of their FTA (ADB 2008). A proposed FTA between China and the ASEAN members would open further opportunities to integrate Toray and Itochu textile operations across the region. New channels may permit better flows of human resources across national borders, as well as of materials, products, and ideas.

Thailand: coordination without Competition

Local firms appear prominent among spinning and weaving mills in Thailand, though not upstream in man-made fibers where firms from Taiwan, India, and Japan jostle for market share. But if competitive with foreign mills midstream, local firms have found their niche downstream in apparel. Growth in foreign-invested firms noted above has advanced Thai industry from electrical machinery to more sophisticated electronics, and from local to export markets in automobiles. A similar trend can be seen in the rapid expansion of man-made fiber production over the past decade among foreign invested firms. Beyond simply expansion of capacity, the availability of more sophisticated fibers and fabrics has enabled apparel firms to move up the fashion chain, a second significant trend in the industry.

Nearly a decade ago some doubted the competitiveness of Thai apparel exports in the face of China's lower-cost garments (FTI 2008). The Thai industry has not only survived but expanded, holding and expanding foreign investment (ATB 2005). Mills and merchants have survived the China threat so far by providing higher-quality goods. Japan's GTCs such as Itochu and Marubeni pioneered Thai apparel exports and remain investors. Marubeni has played a more prominent role than Itochu in leadership of the main local Nikkei associations, the JCCI in Bangkok. Marubeni executives held the prestigious presidency of JCCI six different times from 1971, and as recently as 2006. Marubeni's Tetsuji Banno serves today as one of three vice-presidents (JCCI-B 2008).

Recent news of Marubeni's withdrawal from its leading Thai textile investment, Erawan Textile Company, struck an ominous note for the industry (BP 2008b). China's looming presence adds to the urgency of extending knowledge networks in Thai textiles to anchor production. The recent JTEPA details new forms of coordination between the industries of the two nations, despite the absence of competitive local counterparts to upstream Nikkei textile networks with deep roots in Thailand. A nation which has anchored investment over the past decade now faces the challenge of extending knowledge flows to survive the China challenge.

Sector

Centrality in the domestic economy, growth, and trade integration with her neighbors help to distinguish the industry. Textiles and apparel registered a 10.7 percent share of GDP in manufacturing in 2005, and one-fifth of employment in manufacturing, The sector registers the largest number of firms in manufacturing, with 2,528 in apparel alone, apart from upstream sectors (THTI 2007). Toray and Teijin and other foreign firms operate a full line of mills stretching from man-made fibers down to the weaving of fabrics, and some to garments. Local manufacture of chemicals such as TPA and caprolactam has grown dramatically, but still does not meet local demand. Meanwhile imports of ethylene glycol have grown (THTI 2007: 3.23). Officials have eliminated tariffs on imports of other major raw materials for synthetics fibers, to support upstream production (KCFA 2008: 161). Spinning mills must also import cotton and wool from abroad, and then blend with man-made filaments for polyester-cotton blended fabric.

Dependence on foreign-invested firms has not stunted growth since recovery from the Asian Financial Crisis. Production of man-made fibers jumped from 806,000 tons to 848,000 tons in the same years (THTI 2007: 2.5). Thai Toray Synthetics has installed capacity for 27,000 tons of nylon filament annually, the largest nylon mill in Thailand. But polyester dominates synthetic manufacturing in the nation, with Teijin Polyester and Teijin Thailand alone listing installed capacity of over 230,000 tons annually. Toray Synthetic in Ayutthaya registers capacity of nearly 40,000 tons annually of polyester filament (KCFA 2008: 200).

Production of higher-value-added knitted fabrics rose from 251,000 tons to 288,000 tons (THTI 2007: 2.8). Annual exports grew from \$5.1 billion in 2002 to \$6.3 billion in 2006, nearly 5 percent of total Thai commodity exports (Table 7.4). Apparel exports of nearly \$4 billion dollars in that year overshadowed \$2.3 billion of synthetics and fabrics. Synthetics and fabric dominated South Korean exports, while apparel exports far outpace textiles in China's export profile. We find the Thai industry in transition upstream towards higher-value-added exports of textiles, yet maintaining slower growth in apparel.

Table 7.4 Value of Thailand's trade in textiles and apparel, 2006 ($US millions)

Commodity/Trade	*Imports*	*Exports*
Synthetics (A)	754.5	1,566.0
Fabrics (B)	885.4	852.1
Apparel (C)	322.7	3,901.2
Total textiles (A+B+C)	1,962.7	6,319.5
Total Thai commodity trade	128,584.4	$130,580.0
Textiles/Total trade	1.53%	4.84%

Source: UN (2008).

* HS 2002 SITC Code 54–63. I include in the category of synthetics man-made filaments (Code 54) and man-made fibers (Code 55). In the category of fabrics, I include wadding (56), carpets (57), wovens (58), coated fabrics (59), and knits (60). In the category of apparel, I include knitted apparel (61), general apparel (62), and other made-up textile articles (63). My category of textiles includes amounts from all ten codes.

Table 7.5 Value of Thailand's trade in textiles and apparel with China and Japan, 2006 ($US millions)

Commodity/Trade	*Import from China*	*Export to China*	*Import from Japan*	*Export to Japan*
Synthetics (A)	182.8	144.7	128.8	43.7
Fabrics (B)	246.7	57.2	91.8	78.2
Apparel (C)	118.6	12.1	24.6	246.3
Total textiles (A+B+C)	548.2	214.1	245.4	368.2
Total commodity trade with partner	13,617.1	11,774.1	25,842.6	16,542.4
Textiles/Total trade with partner	4%	1.8%	0.9%	2.2%

Source: UN (2008).

* HS 2002 SITC Code 54–63. I include in the category of synthetics man-made filaments (Code 54) and man-made fibers (Code 55). In the category of fabrics, I include wadding (56), carpets (57), wovens (58), coated fabrics (59), and knits (60). In the category of apparel, I include knitted apparel (61), general apparel (62), and other made-up textile articles (63). My category of textiles includes amounts from all ten codes.

Trade directions also distinguish a Thai node in global textile value chains from its neighbors to the north (Table 7.5). Textile imports from China account for only 4 percent of total Thai imports from its neighbor, with fabrics and synthetics outpacing apparel imports. Textiles and apparel account for 23 percent of Japan's imports from China, and nearly 9 percent of South Korea's China imports. Locally made garments still compete with Chinese imports in the Thai market. South Korea and especially Japan are closely linked to the Chinese textiles in both trade and investment, in contrast to Thailand's relative distance. Perhaps more intriguing is the relatively small role that Japan plays in the Thai textile trade. Re-imports link Japan to the textile industry in China, with Nikkei firms processing Japan-made fabrics into fashion which is then sent back to the home islands. Geographical distance and higher costs may explain why Nikkei merchants and mills with long histories in Thailand, such as Itochu and Toray, have chosen third-country exports (ATB 2005). One benefit for the Thai industry is stronger linkages to GVCs beyond Japan.

Nikkei networks

An Itochu representative served as president of the JCCI in Bangkok already in 1957 (JCCI-B 2008). Itochu (Thailand) dates from 1974 and continues today with holdings in food production, specialty chemicals, and textiles. The firm holds 14 percent of the shares in a cotton-spinning venture with Japan's Shikibo, employing about 250 workers (TKS 2007b). TTL Industries also appears on the Itochu website, a spinning and weaving venture in yarns and fabrics, where it maintains a minority share. If Marubeni is withdrawing due to competition from lower-cost competitors, Itochu appears to be holding for the time being but not expanding, perhaps to integrate with its China-based textile enterprise with further trade liberalization.

In contrast, Toray has reinvented itself as a chemical as well as a textile maker in Thailand. Toray points to Thai Toray Textile Mills in 1963 as its first offshore production venture, followed by Toray Nylon a few months later. Thai Toray Textile Mills continues today with 600 workers spinning and weaving nylon and polyester filament. Headquarters decided to extend the production line to fabrics a decade later as majority partner with local investors in Luckytex, a 51:49 joint-venture. Luckytex provides airbags for the local auto industry using nylon thread from nearby Toray plants, employing 2,300 workers. But the core of Toray investments remains what was originally Toray Nylon. The latter was reorganized in 2003 as the wholly owned Thai Toray Synthetics, including also newer investments such as Thai Toray Fibers, and 3TM Plastics Company. Now operating as a general chemical firm, the enterprise produces nylon, polyester filament, and engineering plastic compounds with 760 employees, including sixteen Japanese engineers (ATB 2004; TKS 2007b). Moving further up the chemical production stream recently, Toray and Mitsui Chemicals joined in 2004 to establish Thai Pet Resin. The lead partners each hold 40 percent of the shares, together with the local Cementhai holding the remaining 20 percent.

Knowledge flows have been at the center of Toray enterprise in Thailand for nearly five decades. Toray Chairman Hiroshi Maeda spoke of globalization and technology transfer in an address to the Japan–Thailand Economic Cooperation Society (JTECS) in Bangkok in 1996. Incremental, continuous development was his theme, evident in the chronicle of expanding Toray investment from Toray Nylon and Textile Mills in 1963 to Luckytex in 1972, and then 3TM Plastics and Toray Fibers a decade later. Global trading arrangements in textiles such as the quota system under the General Agreement on Tariffs and Trade (GATT) prompted

Toray to move downstream in Thailand. Yet growth in the local economy and industry prompted more sophisticated production, such as polyester filaments and base film (Maeda 1996a). Maeda then turned to Luckytex as a case study of technology transfer, citing reinvestment with technology-intensive air-jet looms and training programs. What resulted was a shift away from labor-intensive, dangerous working conditions towards greater local expertise and operating competence. Evidence included declining employment from 6,700 in 1980 to only 2,500 in 1996, yet rising production levels. Coupled with better technology and expertise was a remarkable improvement in safety at the plants (Maeda 1996b).

We noted above the organization of JTECS on the Japan side in 1972, and Thai counterpart, the Technology Promotion Association (TPA) the following year, and their recent evolution into a university program supported by Toyota. Though less prominent in the university project, Toray has been a central force in promoting the initial organizations since their inception. Toray's textile and chemical operations have grown with the apparel sector, chemicals, and now also the automobile industry in Thailand. Knowledge exchange has deepened at each stage. The firm has been prominent in promoting engineering education and wider training for technology workers as well. A former Toray CEO, Katsunosuke Maeda, serves today as chair of JTECS, continuing the firm's decades of support for the organization (JTECS 2007). JTECS' goals include "contribution to the economic and technological development of Thailand" through human resource development and technology transfer programs.

Seminars and publications on the Japan side help to sustain corporate and cultural interest, but the more important efforts take place on site at TPA properties in Bangkok with over 250 staff. Activities in the first six months of 2006 included language programs for 10,000 students, management training for 40,000 students, and 19,000 consultations for calibration services. TPA makes clear its goal of disseminating "Japanese management systems such as Toyota's production system that can be said to be a symbol of Japan's manufacturing skills (TNI 2007: 4)." With the addition of the TPA-affiliated Technology Promotion Institute (TPI) from 1996, "TPA has become of stronghold of Japan–Thailand economic and technical cooperation (TNI 2007: 4)."

Complementarity

Nikkei textile interests played a major role in designing and promoting the partnership agreement between Japan and Thailand. Toray and Teijin executives served among four private sector representatives from Japan in the initial JTEPA Task Force Meeting, in May 2003. Two executives from Teijin and one from Toray served among five corporate representatives in the next meeting, in December 2003 (MOFA 2003). When negotiations concluded, Prime Ministers Abe of Japan and Surayod of Thailand announced seven cooperation projects in their April 2007 Joint Statement on JTEPA. We looked above to the "Automotive Human Resource Development Institute," and here to the "Textile and Apparels Cooperation" project (Abe and Chulanont 2007).

Goals of the cooperation project include better entrée to the Japanese apparel market, Japanese support for fashion and fabric design, and investment in better technology by Japanese textile firms already on site in Thailand. The single most important new initiative is the Textile and Apparel Research and Design Center. Japan would provide advice on structuring the Center, personnel for training, and the necessary equipment for textile pilot plants. But even here, the Japanese contribution continues the long tradition of dispatch of Japanese engineers and technicians to Thai firms and training centers, such as the Thailand Textile Institute (Jirapet and Amari 2007b).

The partnership brings into focus long efforts to build complementarity in standards, associations, and supporting industries for the Thai textile industry. The Thai Chemical Fibers Manufacturers Association is perhaps the best example of continuity between the industries of the two nations. Teijin, Toray, and a local firm called Asia Fiber Company inaugurated the Thai Synthetic Fiber Manufacturers' Association (TSMA) in 1976. Among the eight polyester firms in the Association today, two belong to Teijin and one to Toray, and the others to Taiwan and Indian firms. Toray also appears among the five members of the Nylon Group. Members have agreed to priorities such as supporting the Thailand Textile Institute, improving the TSMA database, and persuading all synthetic fiber producers to join (TSMA 2008). Yet the small number of local firms in man-made fibers, or indeed among quality mills with integrated textile production lines, makes this cooperation through associations quite different from the competitive environment of South Korea or China.

Prospect

A new Japan–Thai partnership in textiles and apparel holds out little immediate promise for Thai penetration of Japanese apparel markets. Thai firms cannot compete in price with Chinese products in Japan's import market. Moreover, Japanese importers have little reason to switch from the well-articulated Nikkei networks in China for re-imports, to imports from unrelated Thai firms at a greater distance. A more likely scenario would be closer links between SMEs of the two nations, with Japanese SMEs expanding investment in Thai production. The Thai call for Nikkei expertise in fashion and fabric in JTEPA may best be met with "cooperative learning" between the more sophisticated and specialized Japanese firms and Thai "supporting industries." Few small firms can specialize in advanced technologies, or command the research resources necessary to constantly move ahead on the technology frontier. Of greater interest here is the ability of SMEs to transfer "tacit knowledge" of production and organization abroad to maintain and enhance their flexible response to new market opportunities at home and abroad.

Thailand's place within trade flows will probably sustain investment among textile moguls such as Toray and Teijin, and renewed efforts to develop technicians as investments continue to expand. Toray has nurtured an extensive network of synthetics production sites across ASEAN, as well as chemical plants for raw materials, and fabric sites. Liberalization of trade flows within ASEAN permits better specialization in placement of technology, and stronger flows of people and ideas. Thailand's FTA with China promises still more, enabling Toray to balance product, supply and demand across a broad range of production sites. Thailand will not anchor the R&D for Toray in the region, but securing continuing investment as one of multiple hubs will still permit deeper knowledge flows.

Conclusion

Capturing a place for local minds in the flow of global designs across textile and fashion value chains depends in part on place, in part on space. A few South Korean firms such as Kolon have begun to penetrate a space of product and fashion brands, moving from sub-contract manufacture (OEM) of foreign brands to manufacture of their own brand (OBM). China's Youngor has tested the same waters, but now in alliance with global leaders such as Itochu and Marzotto. The Fashion Knowledge Creation Center at the Thailand Textile Institute, together with annual fashion exhibitions in Bangkok reflect a similar aspiration in Thailand. A combination of production site and a situation insuring access to ideas, expertise, and codified information would contribute to a shared time and space where knowledge could be shared and possibly created.

Thai officials did well to insist on Japanese support for their fashion enterprise in the JTEPA negotiations, but now must gain Japanese private sector participation to implement the agreement. Having major mills such as Toray, Teijin, and Toyobo in place on Thai soil has helped to bring technology and ideas to the local industry to date, although without partnership with comparable local firms. The task of advancing upscale in apparel poses new challenges. Nikkei networks in South Korea and China partnering with strong local firms, may prove more congenial sites for developing human resources for design and manufacturing of higher-value-added apparel.

Chemicals and industrial textiles offer a similarly appealing path of growth for South Korea and China, extending the technological strengths of Nikkei producers already in place. South Korea has advanced in both areas, though Toray Saehan remains the leader in industrial textiles. Cooperation with the electronics giant Samsung highlights the cross-sectoral benefits of having a Nikkei leader such as Toray producing on the ground in Korea. Toray Saehan in turn learns more of applications through interaction with Samsung. Beyond simply Nikkei networks on the peninsula, the development of South Korean chemical, textile and apparel networks in China offers new opportunities for coordination between the industries of South Korea and Japan. Toray and Saehan have already turned the page on this story, with Toray Saehan producing spunbond at the Toray China complex in Ningbo.

A mix of foreign and local firms has greatly expanded chemical production in China, providing local sources of raw materials for man-made fibers. But restructuring China's sprawling textile and apparel industry remains a central task for the Chinese government. Apparel exports represent a significant portion of China's trade, but depend in part on low cost production in a country with rising wage levels. Improving local fabrics would permit higher-value-added apparel.

Reinvestment of revenues in a complex of man-made fiber mills has resulted in tremendous capacity for polyester production. Integrating upstream fiber into higher-quality midstream fabric remains a challenge. Nikkei merchants and makers have allied with some local partners, providing a flow of expertise and technology for product development and applications. Local partnerships may be the key to more extensive flows, with local firms such as Youngor, or Weiqiao, gaining a place in Nikkei flows of manufacture and design.

Beyond firms, complementarities in social systems of textile production continue to advance in the region. Concentration of production facilities, as well as extension of production networks has helped to create common interests in balancing supply and demand. Our review of Chemical Fiber Associations in the four nations suggests strong continuities. Flows of information across associations can support new flows of interest in state policy directions across national borders. Cooperation in ACFIF may prove a viable path for sector-specific regionalism.

8 Mapping knowledge networks

Over a decade has passed since scholars chronicled the Asian shift from flying geese to swarming sparrows (Hobday 1995; Moon 2002). An orderly ladder of Japanese technology transfer in the region was giving way to a new technology competition. Mapping the same landscape among production networks today, I find that South Korea has kept pace, with China in hot pursuit, and Japan still up front. What does the future hold for the region? A shift from offshore plants to integrated knowledge networks would help to secure Japan's role, improve human resources and advance technology in host countries, and to strengthen regional integration. South Korea faces a similar challenge but without the breadth or depth of Nikkei networks at present. China may well be ready for the challenge and for industry moves abroad in the coming years.

But Japan finds itself at a critical juncture for upgrading its extensive production lines. Domestic debates over regional leadership, the China threat, and benefits of trade agreements have challenged the merits of Japan's "soft power," or "leadership from behind." Knowledge networks represent a new type of regional leadership with benefits for home and host in manufacturing investment abroad, and for the nascent East Asian Community.

Advances in technology help to define the "New Asia," but other economic and political factors have marked the transition also over the past two decades. Trade in intermediate goods has deepened interdependence. China's ascendance as a manufacturing and trading powerhouse reshaped trade directions, creating an alternate center to Japan, though with a far different profile of products and technology. On the investment side, China captures a growing share of Japanese and especially South Korean outward FDI. Korean investors have pushed into China and Southeast Asia, deepening inter-regional ties and competing with Nikkei networks. Trade and investment both promote and reflect extension of production networks in the region, linking economies in new ways across national borders.

Apart from practice, the process of trade liberalization has fostered FDI, and reinforced production networks. ASEAN + 1 trade agreements with China, Japan, and South Korea promise further integration. Perhaps the most significant feature of the New Asia is neither technology nor trade, but rather state-led efforts to shape an East Asian Community. I look here to the future of networks and knowledge in East Asia, and conclude with Japan's role in the regional community

Networks

I began with "communities of practice" created by actors cooperating for mutual advantage (Fruin 1998). Case studies of Nikkei firms gave form and texture to communities of practice in East Asia. But the shift in offshore networks from strong ties within "company as community" to weak ties among a variety of local sub-contractors, more fluid labor markets and diverse institutional actors in communities of practice changes the boundaries for tacit knowledge (Whittaker and Cole 2006). Management in high-technology firms must harness research resources across the world – a formidable challenge to the solidarities and trust typical in the more insular company as community. Rather than focus on knowledge in "knowledge firms" of high-technology sectors, I looked to mainstream manufacturing lines abroad in neighboring Asian nations to assess potential for incorporating local knowledge resources.

Craft and concept

Altenburg and colleagues admitted their emphasis on innovation over simply "catch-up" for India and China was indeed ambitious (Altenburg, *et al.* 2008). Corporate research and development includes basic research, applied research, and product development. *Innovation* might occur in any of the three steps, but some reserve the term "innovation" for discoveries in basic and applied research. I found a clear division of labor with the first two steps largely at headquarters, and the latter divided between home and host production sites. Hobday had earlier focused on a cumulative process of incremental learning in East Asia, parting ways with those advocating a "leapfrog" approach to technological advance in the race for economic development.

I found evidence of a shift from production to knowledge networks with broader streams of craft and concept. Toyota provides its own examples of craft in subcontracting where SMEs tasked with continuous improvement can specialize in particular parts. This permits assemblers to focus on "developing the product concept and system designs, high value-added parts such as engines and bodies, and strategic issues such as product planning and international strategy (Osono 2006a: 28)."

A review of national innovation systems brought attention to infrastructures in the social systems of production supporting basic and applied research relevant for manufacture. The relative contribution of knowledge networks to regional integration will rest largely on the choice of open versus bounded innovation systems. Our study examined boundaries within a society of sectors. Diversity of innovation systems highlighted diversity of social systems of production. Nikkei networks, including both state and the private sector, have worked over time to develop complementarities in the political economies of host nations. Analysis of specific sectors whether electronics, automobiless, or textiles brought to the fore critical differences in state policy on IFDI and technology sharing in the organization, independence, and activities of industry associations, and in the landscape of local corporate scale, ownership and activity. Historical precedents of "cooperative learning" in inter-firm networks offered a model of the interplay of craft and concept (Okada 2001).

Practice

Localization efforts at Panasonic, Toyota and Toray revealed a shift from vertical integration between headquarters at home and plants abroad, to horizontal coordination with regional headquarters closer to local expertise in labor and consumer markets, and local state policy. Denso pulled us deeper into the craft of making things (*monozukuri*) with its multiple tiers of sub-contractors, and into the regionalization of networks, whether with movement of products within ASEAN, or with a South Korean subsidiary producing in China. Itochu drew us outward between manufacture and market with the processing trade in fashion. The trading firm stretched our focus from regionalization with Prominent Apparel in Hong Kong or Itochu Textile in Shanghai, to global networks in the tripartite joining of Italian and Chinese apparel giants.

Firms coordinate a stream of codified and tacit knowledge through learning on the job, training programs, and dispatch of personnel. Knowledge is embedded in the use of technology. Beyond initial investments, upgrading machinery and product in plants abroad brings new skills to local personnel. Formal training programs as well as manuals within firm, association, and institutes fostered a transfer of codified knowledge. Dispatch of Japanese personnel for on-site calibration of machinery, training, and problem-solving would permit transfer of tacit and codified knowledge. For instance, Koike highlighted the contribution of well-trained Japanese inspectors to sub-contractors in China's apparel sector (1998). But what of local knowledge?

Local learning

Prosperous consumer markets in the New Asia have opened new opportunities for Nikkei manufacturing abroad. Panasonic has tapped local expertise in China's consumer markets, and manufacturing expertise in bringing new applications to market. Toray Saehan and Denso's Poong Sung joint-venture have tapped South Korea's expertise in design manufacture. Toyota's International Multi-purpose Vehicle (IMV) Project in Thailand highlights the challenge of deepening knowledge flows in East Asia. Toyota Motor Thailand president Ryoichi Sasaki wrote in 2005 of a transfer of know-how, exchange of personnel, and a flow of parts. Beyond flows of codified and tacit knowledge along these dimensions, Toyota officials wrestled with the question of "the right proportion of design capability to transfer to the Center and when (Osono 2006b: 20)."

Precedents of cooperative learning in Japanese inter-firm networks suggest prospects for a cross-border pyramid of supporting industries for Japanese firms in Asia. METI recently proposed a network of offshore local SMEs in support of larger firms (METI 2007b). Finding a substantive role for Japanese and host country SMEs appears critical for extension of knowledge flows in the region. Small firms represent the majority of enterprise and employment in the three sectors of interest in China, South Korea, and Thailand. Advantages of corporate mobility and geographical proximity, coupled with the demand for low-cost production sites took small firms abroad. SMEs from Japan and South Korea pioneered FDI in China, and Japanese SMEs played that same role in Southeast Asia.

Small firms would cooperate with larger firms to learn specifications and necessary programs of manufacture, and then specialize in a certain aspect of product or process. Effective technology support institutions, whether at firm, association or school, appeared critical for learning. Further research on support institutions abroad might clarify the feasibility of such a model. Apart from support systems, any sharing of technology or expertise demands trust. Complementarity with Nikkei social systems of production appears a necessary condition for establishing trust in inter-firm relations.

Knowledge

Knowledge networks can be understood as interactive linkages around nodes of tacit and codified knowledge embedded in GVCs. Firms within sectors help to embed the social practices critical for effective economic coordination and create nodes (Storper 2000). Firms such as Panasonic, Toyota, and Toray rank as Japan's leaders in consumer electronics, automobiles, and textiles respectively. Itochu leads general trading companies in fabrics and fashion, and Denso stands at the head of the line among auto component firms. Beyond manufacturing routines on the plant floor, firms advance their technology levels with in-house research units, industry associations, and sector-specific technology regimes supported by the state. Manufacturing firms lead worldwide in R&D funding. Panasonic and Toyota rank among Japan's leaders in number of patents. Denso and Toray are research leaders in their own sectors among local and global competitors.

Sectors of electronics, automobiles, and textile and apparel advanced along similar paths and with similar organization in the three nations of Northeast Asia. Thailand has anchored investment in all three industries, although without strong local competitors. State and state FDI policy remain the major discontinuity among the four nations, with some similarity between strong states' insulation in Japan and South Korea, but discontinuity with China's socialist state or with the relatively weak state in Thailand's history of industrialization. Knowledge networks come to the fore at the intersection of firms, sectors and states within GVCs.

Firms

Corporate organization in individual nations helps to channel exchange of information. South Korea's concentration in family-owned conglomerates provides a foil to mainly supporting industries of SMEs in Thailand. China's dominant pattern of large, state-owned enterprise presents a quite different set of challenges for establishing borders in the flow of tacit knowledge. South Korean *jaebeol* in electronics such as Samsung or LG command the capital and technology resources to compete directly with major Japanese firms. A similar pattern is evident in auto assembly where Hyundai competes directly with Toyota. Strategic alliances offer promise for new avenues of knowledge exchange. But Nikkei manufacturing networks in South Korea enjoy other opportunities in sectors where Japanese firms partner with strong local firms.

Denso and Toray have not only tapped local engineering and manufacturing expertise in auto parts and textiles, but extended from a South Korean manufacturing base into Chinese manufacturing subsidiaries such as Poong Sung Electronics in Tianjin and Toray Polytech in Nantong. Partnerships rather than simply alliances, Denso and Toray built upon their manufacturing commitments in South Korea to extend knowledge flows. Bilateral similarities in corporate structure offer a congenial context for knowledge exchange at industry associations or sectoral technology regimes, but manufacturing anchors on the peninsula appear necessary to hold on to extensive knowledge flows among firms.

Corporate structure in Thailand provides a context of cooperation without competition, permitting knowledge flows of craft rather than concept. Thailand has secured extensive manufacturing anchors in Nikkei networks for all three sectors. Tariff reductions within ASEAN, and preferential trade agreements with China and now India will offer Nikkei firms wider opportunities for integrating production within the region from a Thai base. South Korea has been reluctant to anchor Nikkei production until recently, but aggressive in the alliances and bilateral partnerships or joint-ventures that permit deeper flows of tacit knowledge. What has emerged in Thailand instead is an array of supporting industries, more in the pattern of SME networks in Japan. The partnership agreement (JTEPA) with Japan promises to link SMEs networks in the two nations for technology transfer and possibly joint-investment in other parts of ASEAN.

Nikkei firms must adjust to state-owned enterprise in China, where local partners represent a public as well as corporate interest. A few electronics firms in China such as Haier have emerged recently as competitors, but in a limited range of consumer electronics. Still others such as Huawei in telecommunications equipment have joined with European or US firms to gain advanced technology and may well "leapfrog" the conventional catch-up process. Sanyo has partnered with Haier for production in Thailand, and Panasonic continues in a blend of sole ventures and partnerships within China, as well as multiple R&D institutes.

Partnerships have been the rule for foreign investors in auto assembly. Difficulties of state versus market priorities were evident at Toyota's earlier joint-venture with Tianjin Automotive in the mid-1990s. Acquisition of Tianjin Auto by First Auto Works has proved a far better match for Toyota, with multiple models bringing new designs to the market. Toyota appears better positioned now to tap local engineering resources for both craft and concept through both its research facilities and assembly lines. In contrast, Toray entered the China market with largely sole ownership, and only later engaged extensively with local partners for applications. For instance, Toray has allied now with successful spinning and weaving firms in China to extend its production into blended yarn and fabric. Similar to Panasonic,

Toray needs local talent to identify and help adapt manufacturing technology for local product applications. Itochu pioneered bilateral knowledge flows, initially bringing Japanese craft in apparel to Chinese plants, and now brokers multi-lateral knowledge flows in fashion tapping local expertise in large-scale manufacturing and marketing. Japan faces a new competition in China for human resources with local and foreign firms, particularly evident in the electronics sector.

Sector and local development

Contrasts between modular and integral production can be very significant for local development. Modular production in electronics with standardized interfaces brings employment and investment to the host nation, but weak links to knowledge flows beyond basic assembly techniques. Nonetheless, state and private sector combined in South Korea to move up the production ladder through licensing and technology agreements to create a local electronics industry. China has likewise begun to advance, but through a blend of local and foreign-invested firms. Thailand has succeeded in anchoring Nikkei electronics manufacturing, and indeed, advanced with the firms from basic consumer electronics into semi-conductors. Chinese and South Korean industries have both permitted import-substitution and now provide exports as well. Nikkei networks in China have begun to tap local knowledge resources through research institutes and joint-ventures. Alliances appear to be the most viable bilateral cooperation with South Korea to date. The foreign-invested Thai industry serves the local market as well as driving a huge export stream, but again without comparable local partners to push a knowledge exchange.

Integral production in automotive manufacturing and in downstream apparel and fashion remains the strength of Japanese industry, and an opportunity for local development. South Korea has long anchored Nikkei production in auto components and remains tied to Japanese technology streams in higher-value-added parts. At the same time, the industry has succeeded in developing its own design and manufacturing capabilities, moving well beyond import substitution to heated competition with Japan in foreign markets. Investment by General Motors and Renault in Korean auto production has changed the sectoral landscape and added a new competitive dynamic to East Asian knowledge networks in a critical sector. Similarly, the investment by Shanghai Automotive in Korea's Ssangyong Motors may provide precedent for new bilateral channels of craft and concept with China.

While many laud South Korean success in integral manufacturing, some critics fault weak craft and reliance on foreign technology in Chinese manufacturing as evidence of shallow roots in GVCs (Lemoine and Ūnal-Kesenci 2004; Ozaki 2007). Localizing a foreign-invested industry in China will depend in large part on effective flows of both craft and concept. Nikkei firms can increase their market share with local partners only by expanding a local base of talent and using sub-contractors to increase production of more sophisticated vehicles. Concerns about intellectual property rights may border flows of tacit knowledge in areas of higher-value-added production. Thailand has developed better protections of intellectual property, and has gained craft in supporting industries. Benefits include employment, training, and investment, and a thriving local consumer market for autos and pick-up trucks. Two recent shifts in the sector have further served Thai development goals. A turn from domestic to export markets has spurred rapid expansion of auto manufacturing. Second, regional integration has spurred development of Thailand as the automotive hub in Southeast Asia.

Changing borders among sectors have likewise affected the anchoring of Nikkei manufacturing in East Asia. Panasonic provides consumer electronics for the automotive industry in China and Thailand, supplying Nikkei assemblers. Toray and Itochu supply automotive interiors to Nikkei assemblers in the same two nations. Thailand has reinforced production anchors by the triple nesting of manufacturing in Nikkei electronics, auto and textile industries. Efforts of the Japanese state and peak industry associations to build complementarity in local industry associations, trade policy, and property rights can yield benefits beyond the interests of simply one sector. Multiple major firms at the JCCI in Bangkok can mobilize a common industry voice on major issues of market and manufacture. South Korea provides the foil, with weak nesting of Japanese manufacturing in electronics and auto assembly, despite continuities in the interests of similarly structured industry associations. China has succeeded in nesting extensive Nikkei production in all three sectors recently, although Itochu and Panasonic put down roots much earlier than Toyota and Toray.

Mapping community

The term *community* refers to bounded networks with an emphasis on interactions rather than individuals or social systems (Tilly 1998). Interactive, bounded networks might include such diverse interactive groups as trading systems, military hierarchies, or even job-seeking networks and all fall within the term "community." Of particular interest is the interplay of communities or bounded networks, and nation-states. Trans-state and interstate communities have long moved across the relatively recent borders of nation-states, particularly in Asia with its strong ethnic networks. Among the most significant of transnational networks in the Asia–Pacific region today is the network of production lines which fuels increasing volumes of regional commerce. We can characterize Asian integration with production networks or an economic driver from below, rather than by politically negotiated and constructed regionalism from above (Hurrell 1995). Value chains or production-sharing networks, then, can be viewed as an exchanging or shaping community within the region promoting commerce. The dual emphasis on geographical location as well as flows across space of knowledge also permits a dual focus on the "value-added" production important to investing firms seeking profits on their technology, and to "development" among local firms hoping to gain a place in the network.

Pempel wrote recently of the "remapping" of East Asia to highlight new lines of cooperation that redefine our conception of the region (2005). He argued that cooperation complicated and often competed with traditional lines of sovereign state power. Others have questioned the prospect of East Asian solidarity, noting the absence of common values or positive experiences of cross-border exchange, resentments over historical aggression, and residual fears of expansionism in the region. Nonetheless, the East Asian Community continues to evolve with annual summits, a web of bilateral trade agreements, and the ASEAN + 1 trade pacts with China, Japan, and South Korea. I return in conclusion to the question of how the remapping efforts of firms and states might affect techno-nationalism in the region.

Standards

Globalization has pushed firms abroad for markets and lower-cost production, and now for local knowledge resources as well. Kikkawa's contrast of global operations at Toray versus global products at Sanyo appears quite dated now almost a decade later (2000). Each of the firms in this study has moved aggressively to position production near markets, and to tap regional knowledge sources. Equally significant is the growing cooperation at the sectoral level across borders. Panasonic can promote standardization of specifications across national markets in the APEC Business Advisory Council, in tandem with Toyota representatives seeking similar emission standards. The APEC Automotive Dialogue joins Toyota and Denso with Hyundai and Hyundai Mobis, as well as US auto giants, to harmonize standards in the auto industry. Joint consultation among Chemical Fiber Manufacturers Associations has supported a continuing dialogue on market and technology developments in the ACFIF. Basic tasks of information exchange open opportunities for cooperation on standards, and perhaps in time on developing consumer markets and appropriate technologies.

Yoshimatsu argued that Japanese state recognition of the role of technology standards in industrial competitiveness has now become part of state technology policy (2006). Meanwhile industry leaders in core sectors have begun to cooperate in technology applications in their regional production networks. Japanese officials have reoriented policy to regional rather than simply to national standards, with efforts to diffuse Japanese standards among nations of the region. Although he suggests a transition away from techno-nationalism to "techno-regionalism," the state strategy appears to simply reinforce techno-nationalism with Japanese leadership in standards. Cooperation with regional partners on standards in industry associations and other fora remains a critical element of the seamless market goal promoted by JETRO in the region.

Epistemic community

Japanese officials have made a dramatic effort recently to regain Japanese leadership in the dialogue on community, and to establish an epistemic network of institutes across the region to reinforce production lines. The initiative began with the METI proposal to expand ASEAN + 3 to an ASEAN + 6 framework, joining the ten ASEAN nations not only to China, Japan, and South Korea, but to India,

Australia and New Zealand as well. Japanese leaders made clear their market priority in naming the initiative the "Comprehensive Economic Partnership in East Asia (CEPEA)." Officials mandated CEPEA with the goal of strengthening industrial competitiveness in the region through promoting the formation of efficient production networks (Nishimura 2007a: 158).

Promoting economic integration through regional production networks to this point harnessed a market dynamic from below to official encouragement from above. But Japanese leaders pressed further to establish new networks for a regional exchange of ideas and policies on the model of the OECD in Europe. Officials sweetened the CEPEA proposal with funding to establish research units in each of the member nations in the Economic Research Institute for ASEAN (ERIA) and East Asia. The organization would provide a tripartite-type forum for policy dialogues and interactions among researchers, policymakers and the private sector (Nishimura 2007a: 159). At a preparatory meeting in Singapore, Prime Minister Lee Hsien Loong wrote in support of an ERIA that would focus on "practical research to bring new perspectives and solutions to the challenges of regional production networks (Nishimura 2007b: 77)." ERIA was formally inaugurated in June 2008 at the ASEAN Secretariat to "provide a solid intellectual foundation for cooperation (ERIA 2008)."

Samuel Kim recently tempered the enthusiasm of regionalism experts with the sobering fact that investment and trade has not led to greater geo-political security in the region: "Northeast Asian states must now worry not only about their military power but, more important, about the economic, cultural, and knowledge power required to survive and prosper in an increasingly globalized and competitive world (Kim 2006: 169)." ERIA would lend credence and a wider academic voice to the Japanese alternative of ASEAN + 6 through ERIA, an expanded Asian community with a stronger voice for Japan's liberal democratic values due t addition of India, Australia and New Zealand. Whether the political priori undermine the economic focus is unclear at this point. What is clear is the e to regain the initiative in the New Asia among firms and state leaders. Issues of techno-nationalism may yet prove divisive, in Japan as well as in production partners in the region. But cooperation is critical, and cooperation in a shift from production to knowledge sharing ideal.

Notes

1 Business, knowledge, and networks

1 Matsushita Electric Industrial Company (Matsushita Denki Sangyō Kabushiki Kaisha) represented the Matsushita family of firms, including major subsidiaries such as Matsushita Electric Works (MEW, Matsushita Denkō Kabushiki Kaisha). MEI changed its name in October 2008, to Panasonic Corporation. MEW revised its title to Panasonic Electric Works. The offshore subsidiaries of the two publicly traded firms are listed separately, although the two firms have agreed to collaborate on research and development since 2004. I will use the term Panasonic to refer to the combined research activities of the Panasonic Group, and to the offshore subsidiaries of the mother company formerly termed MEI. I will use the term Panasonic Electric Works (PEW) to refer to its separate list of subsidiaries abroad.

3 Commerce and the East Asian community

1 As signaled in Chapter 1, the four newly industrializing economies (NIEs) in East Asia are Hong Kong, Korea, Singapore and Taiwan. Japan's annual trade with Korea is far greater than that with Singapore, but Singapore holds a 3 percent share of Japan's total direct investment assets, compared to only 2.1 percent for Korea (BOJ 2006). Among the ten members of the Association of Southeast Asian Nations (ASEAN), the ASEAN Four refers to Indonesia, Malaysia, the Philippines, and Thailand.

2 UNCTAD (2008b) offers the following definition of FDI flows. "For associates and subsidiaries, FDI flows consist of the net sales of shares and loans (including non-cash acquisitions made against equipment, manufacturing rights, etc.) to the parent company plus the parent firm's share of the affiliate's reinvested earnings plus total net intra-company loans (short- and long-term) provided by the parent company. For branches, FDI flows consist of the increase in reinvested earnings plus the net increase in funds received from the foreign direct investor. FDI flows with a negative sign (reverse flows) indicate that at least one of the components in the above definition is negative and not offset by positive amounts of the remaining components."

3 UNCTAD (2008c) offers the following definition of FDI stocks. "For associate and subsidiary enterprises, FDI stock is the value of the share of their capital and reserves (including retained profits) attributable to the parent enterprise (this is equal to total assets minus total liabilities), plus the net indebtedness of the associate or subsidiary to the parent firm. For branches, it is the value of fixed assets and the value of current assets and investments, excluding amounts due from the parent, less liabilities to third parties."

4 Hong Kong and the Virgin Islands actually rank ahead of Korea among investing nations in the China market, but are excluded here due to the difficulty of identifying the actual origins of their FDI. Investors from the West and Taiwan often use Hong Kong intermediaries to invest in China.

5 Foreign direct investment (FDI) is defined as an investment involving a long-term relationship and reflecting a lasting interest in and control by a resident entity in one economy (foreign direct investor or parent enterprise) of an enterprise resident in a different economy (FDI enterprise or affiliate enterprise or foreign affiliate). Such investment involves both the initial transaction between the two entities and all subsequent transactions between them and among foreign affiliates. A direct investment enterprise is defined as an incorporated or unincorporated enterprise in which the direct investor, resident in another economy, owns 10 percent or more of the ordinary shares of voting power (or the equivalent). However, this criterion is not strictly observed by all countries reporting (UNCTAD 2008a).

6 The Bank of Japan (2007b) reported $9.2 billion of accumulated Japanese investment in Korean securities and bonds in 2006, as compared to only $1.2 billion of Japanese portfolio investment in Thailand.

5 Electronics sector

1 Data provided on the Huawei website, www.huawei.com, accessed June 3, 2008.

References

Abe, Shinzo, and Surayud Chulanont. 2007. "Joint Statement on the Japan–Thailand Economic Partnership Agreement (JTEPA), Tokyo, April 3." Edited by Ministry of Foreign Affairs (MFA). Bangkok: Ministry of Foreign Affairs, Kingdom of Thailand. Online available: <http://www.mfa.go.th/jtepa/asset/joint_statement_jtepa_eng.pdf> (accessed December 19, 2008).

ACFIF. 2006. "Ajia Hwaseom Saneob eui Vision: Juyou Naeyong Yoyag [Vision for the Asian Chemical Fiber Industry: Summary]." Edited by KCFA. Seoul: Korea Chemical Fibers Association (KCFA).

Adams, F. Gerard. 2006. *East Asia, Globalization, and the New Economy*. New York: Routledge.

ADB, Asian Development Bank. 2006. *Key Indicators 2006: Measuring Policy Effectiveness in Health and Education: Regional Tables II – Education Indicators*. Manila: Asian Development Bank.

——. 2008. "Trade and Investment: People's Republic of China–Thailand Free Trade Agreement (Under Implementation)." Manila: Asia Regional Integration Center, Asian Development Bank (ADB). Online available: <http://aric.adb.org/fta.php?id=65&ssid=3&title=People%27s%20Republic%20of%20China-Thailand%20Free%20Trade%20Agreement> (accessed December 12, 2008).

Adler, Paul S., and Charles Heckscher. 2006. "Towards Collaborative Community." See pp. 11–05 in *The Firm as a Collaborative Community: Reconstructing Trust in the Knowledge Economy*. Oxford: Oxford University Press.

Altenburg, Tilman. 2006. "Opportunities for Asian Countries to Catch Up with Knowledge-based Competition." See pp. 21–53 in *Asia's Innovation Systems in Transition*, edited by Bengt-Ake Lundvall, Patarapong Intarakumnerd and Jan Vang-Lauridsen. Northampton, MA: Edward Elgar.

Altenburg, Tilman, Hubert Schmitz, and Andreas Stamm. 2008. "Breakthrough China's and India's Transition from Production to Innovation." *World Development* 36, 2 (February): 325–44.

Anchordoguy, Marie. 2005. "Japan's Technology Policies and Their Limitations." See pp. 189–203 in *Japan and China in the World Political Economy*, edited by Saadia M. Pekkanen and Kellee S. Tsai. New York: Routledge.

Ando, Mitsuyo, Sven W. Arndt, and Fukunari Kimura. 2006. "Production Networks in East Asia: Strategic Behavior by Japanese and US Firms." *JCER Discussion Paper No. 103*: 1–46.

Aoki, Mashiko. 2001. *Toward a Comparative Institutional Analysis*. Cambridge, MA: MIT Press.

APEC. 2007. "General Economic Indicators: Asia Pacific Economic Cooperation."

Singapore: Asia–Pacific Economic Cooperation (APEC) Secretariat. Online available: <http://www.apecsec.org.sg/apec/member_economies/general_economic_indicators.html updated March 3> (accessed October 15, 2007).

Asada, Takashi. 2006. "Becoming a Company that Contributes to the Development of China in the 21st Century: Interview with Takashi Asada, Associate Director, Corporate Management Division for China and Northeast Asia, Matsushita Electric Industrial Co., Ltd." *Erina Report* 12 (November): 8–11.

Asakawa, Kazuhiro. 2007. "Metanational Learning in TFT-LCD Industry: An Organizing Framework." *RIETI Discussion Papers 07-E-029*. Tokyo: Research Institute on Economy, Trade and Industry (RIETI), Japanese Ministry of Economy, Trade and Industry (METI). Online available: <http://www.rieti.go.jp/jp/publications/dp/07e029.pd> (accessed May 7, 2008).

Asheim, Bjørn, and Arne Isaksen. 2000. "Localized Knowledge, Interactive Learning and Innovation: Between Regional Networks and Global Corporations." See pp. 163–98 in *The Networked Firm in a Global World: Small Firms in New Environments*, edited by Erik Vatne and Michael T. Taylor. Burlington, VT: Ashgate.

ASPAC. 2002. "Asia–Pacific Textile and Clothing Industry Forum (ASPAC-TCIF)." Tokyo: Japanese Ministry of Economy, Trade and Industry (METI), Asia–Pacific Textile and Clothing Industry Forum (ASPAC-TCIF). Online available: <http://www.meti.go.jp/aspac/> (accessed October 18, 2008).

ATB, Asian Textile Business. 2004. "Toray Thailand Group: Results of Business Integration at Stake." *Asian Textile Business* (August): 15.

——. 2005. "Heading for Sophistication." *Asian Textile Business*. Osaka: Osaka Senken Ltd, reprinted in redOrbit. Online available: <http://www.redorbit.com/news/science/213244/heading_for_sophistication/> (accessed October 16, 2008).

Back, Hum-Gil. 2005. "The Way Forward for the Korea, China and Japan FTA of Textiles and Apparel." See pp. 1–15 in *Papers of the Fourth Korea–China–Japan Business Forum*, October 13, 2005, edited by FKI. Seoul: Federation of Korean Industries (FKI).

Baek, Yongchun, and Randall Jones. 2005. "Sustaining High Growth through Innovation: Reforming the R&D and Education Systems in Korea." *OECD Working Papers* No. 470. Paris: Economics Department, Organization of Economic Cooperation and Development (OECD).

Beijing Center. 2006. "Chūgoku oyobi Peiking Shūhen Chiiki no Saikin no Keizai Jōtai [Recent Economic Situation of China and the Beijing Surrounding Region]." Beijing: Japan External Trade Organization (JETRO), Beijing Center.

Bell, Daniel. 1973. *The Coming of the Post-industrial Society: A Venture in Social Forecasting*. New York: Basic Books.

BERNAMA, Malaysian National News Agency. 2008. "Oversupply Remains a Critical Factor in Chemical Fiber Industry." Penang, Malaysia: BERNAMA, Malaysian National News Agency. Online available: <http://www.bernama.com/bernama/v3/news_business.php?id=333503> (accessed November 12, 2008).

Bernard, Mitchell, and John Ravenhill. 1995. "Beyond Product Cycles and Flying Geese: Regionalization, Hierarchy, and the Industrialization of East Asia." *World Politics* (January) 47, 2: 171–209.

Biggart, Nicole Woolsey, and Mauro F. Guillén. 1999. "Developing Difference: Social Organization and the Rise of the Auto Industries of South Korea, Taiwan, Spain, and Argentina." *American Sociological Review* 64 (October): 722–47.

BOJ. 2006. "Japan's International Investment Position at Year End 2005." Tokyo: Bank of Japan (BOJ), International Department.

——. 2007a. "Statistics: Balance of Payments, Direct Investment by Region and Industry, End of 2006." Tokyo: Bank of Japan (BOJ).
——. 2007b. "Regional Direct Investment Position and Regional Portfolio Investment Position End of 2006." Tokyo: Bank of Japan (BOJ).
Boontherawara, Apichai. 2007. "Exim Bank Warns of Export 'Illusion'." *Bangkok Post* July 9, 2007. Bangkok: Bangkok Post Company.
Borrus, M. 1999. "Exploiting Asia to Beat Japan: Production Networks and the Comeback of US Electronics." See pp. 213–37 in *Japanese Multinationals in Asia: Regional Operations in Comparative Perspective*, edited by D. Encarnation. Oxford: Oxford University Press.
Borrus, Michael, Dieter Ernst, Stephan Haggard, and University of California Institute on Global Conflict and Cooperation. 2000a. *International Production Networks in Asia: Rivalry or Riches?* New York: Routledge.
Borrus, Michael, Dieter Ernst, and Stephan Haggard. 2000b. "Introduction: Cross-border Production Networks and the Industrial Integration of the Asia–Pacific Region." See pp. 1–30 in *International Production Networks in Asia: Rivalry or Riches*, edited by Michael Borrus, Dieter Ernst and Stephan Haggard. New York: Routledge.
BP, Aranee Jaiimsin. 2007a. "Thailand–Japan Pact Could Damage Auto-parts Industry: Less Local Appeal with Zero Import-tariff Rate." *Bangkok Post* February 26, 2007. Bangkok: Bangkok Post Co. Online available: <http://www.bangkokpost.com/Business/26Feb2007_biz22.php> (accessed February 26, 2007).
BP, Bangkok Post. 2007b. "Exim Bank Warns of Export 'Illusion'." *Bangkok Post* July 9, 2007. Bangkok: Bangkok Post Co. Online available: <http://www.bangkokpost.com/breaking_news/breakingnews.php?id=120021> (accessed July 9, 2007).
BP, Phusadee Arunmas. 2008a. "Thai Exporters Urged to Tap 95% of AFTA Benefits in 2009." *Bangkok Post* November 27, 2008. Bangkok: Bangkok Post Co. Online available: <http://www.bangkokpost.com/271108_Business/27Nov2008_biz38.php> (accessed November 27, 2008).
BP, Krissana Parnsoonthorn. 2008b. "Marubeni Scales Down Textile Business." *Bangkok Post* July 3, 2008. Bangkok: Bangkok Post Company.
Brimble, Peter, and Richard F. Doner. 2007. "University–Industry Linkages and Economic Development: The Case of Thailand." *World Development* 35, 6 (June): 1021–36.
Brimble, Peter, and Shujiro Urata. 2006. "Behavior of Japanese, Western, and Asian MNCs in Thailand: Lessons for Japanese MNCs." See pp. 1–30 in *JCER Discussion Paper No. 105*. Tokyo: Japan Center for Economic Research.
Brown & Co., Ltd. 2001. *Industrial Groupings in Japan: The Changing Face of Keiretsu*. Tokyo: Brown & Co., Ltd.
Brown, Peter, and Rod B. McNaughton. 2002. "Global Competitiveness and Local Networks: A Review of the Literature." See pp. 3–37 in *Global Competition and Local Networks*, edited by Rod B. McNaughton. Burlington, VT: Ashgate.
Camagni, R. 1991. "Local 'Milieu,' Uncertainty and Innovation Networks: Towards a New Dynamic Theory of Economic Space." See pp. 121–42 in *Innovation Networks: Spatial Perspectives*, edited by R. Camagni. London: Belhaven.
Castellani, Davide, and Antonello Zanfei. 2006. *Multinational Firms, Innovation and Productivity*. Northampton, MA: Edward Elgar.
Castells, Manuel. 1996. *The Rise of the Network Society. Volume I, The Information Age: Economy, Society and Culture*. Malden, MA: Blackwell.
CCFA. 2008. "China Chemical Fibers Association: Scope of Activities." Beijing: China Chemical Fibers Association (CCFA). Online available: <http://www.ccfa.com.cn/htdocs/pages.asp?id=13> (accessed November 12, 2008).

CCPIT. 2008. "President Hu Jintao Meets with Japan–China Economic Association Delegation, 10 August 2008." *CCPIT News*. Beijing: China Council for Promotion of International Trade (CCPIT). Online available: <http://english.ccpit.org/Contents/Channel_410/2008/1008/134713/content_134713.htm> (accessed May 15, 2008).

CD. 2005. "Matsushita Invests Heavily in China." *China Daily (CD: N. American Edition)* July 16, 2005 Beijing: China Daily Company.

Chen, Xiangdong, and Guido Reger. 2006. "The Role of Technology in the Investment of German Firms in China." *Technovation* 26, 3 (March): 407–15.

CJCCI (Japanese Chamber of Commerce and Industry in China). 2005. "2005 nen Chūgoku Nihon Shōkai Kaiin Meibo [2005 Membership List of the China Japanese Chamber of Commerce and Industry]." Beijing: China Japanese Chamber of Commerce and Industry.

CNET. 2004. "Samsung, Sony Join Forces on LCDs." Tokyo: CNETAsia.

Cohen, Mark A. 2007. "Current Challenges and Future Prospects in Engaging China on Intellectual Property Rights (IPR) Protection: Some Personal Observations." *RIETI Brown Bag Lunch Seminars*. Tokyo: RIETI (Research Institute of Economy, Trade and Industry). Summary online available: <http://www.rieti.go.jp/en/events/bbl/07112701.html> (accessed May 12, 2008).

Coleman, James Samuel. 1990. *Foundations of Social Theory*. Cambridge, MA: Belknap Press of Harvard University Press.

Corning, Gregory P. 2004. *Japan and the Politics of Techno-globalism*. Armonk, NY: M. E. Sharpe.

COSMOBIC. 2007. "Company Webpage: Partners and Investors." Shanghai: COSMOBIC. Online available: <http://www.cosmobic.com/2004/en/portfolio.htm> (accessed November 12, 2008).

Cyhn, J. W. 2001. "The Political Economy of Technology Development: A Case Study of Korea's Electronics Industry." PhD Dissertation in Division of Social Sciences, University of Oxford.

——. 2002. *Technology Transfer and International Production: The Development of the Electronics Industry in Korea*. Cheltenham, UK: Edward Elgar.

DBJ, Development Bank of Japan. 2006. "Annual Report, Development Bank of Japan, 2006." Tokyo: Development Bank of Japan.

Deffree, Suzanne. 2008. "Panasonic Invests $93.6m in Second China Factory." *Electronic News* 53, 58 (February 4, 2008): 6–7.

Delphi. 2003. "OEM (Original Equipment Manufacture) Business: Delphi Invests in New China and Korea Technical Centers." *Kensei News and Information Service*. Troy Michigan: Kensei News and Information Service. Online available: <http://www.kensei-con.net/BizDev/publish/oem_business/article_18003.shtml> (accessed May 15, 2004).

——. 2008. "Delphi Corporation 10-K For 12/31/07 EX 21." *SEC Information*. Washington, DC: Securities and Exchange Commission. Online available: <http://www.secinfo.com/dsvR3.twj.c.htm> (accessed November 12, 2008).

Dempsey, Eugene. 1996. "Asian Man-made Fibers: The New 'Club of 12' Gears Up to Face the Future – Interview with Katsunosuke Maeda, President, Asian Chemical Fiber Industries Federation (ACFIF)." *ITS Textile Leader* (March): 80–6.

Denso. 2000. "New Investment Strengthens Group's Operations in Korea." *Denso Online Vision Magazine* October 18, 2000. Kariya, Japan: Denso Corporation. Online available: <http://www.densocorp-na.com/newmc/news/newsreleases.asp?ArticleID=73> (accessed May 13, 2002).

——. 2006. "New Company in China to Produce Instrument Clusters." *Denso News* March 15,

2006. Kariya, Japan: Denso Corporation – North America. Online available: <http://www.densocorp-na.com/news/pr.asp?Action=330> (accessed November 12, 2007).

——. 2007. "Denso's Asia and Oceania Regional Headquarters Establishes New Company in Thailand: Technical Center Established as Part of the New Company." In *Denso Global – News Releases* April 3, 2007. Kariya, Japan: Denso Corporation. Online available: <http://www.globaldenso.com/en/newsreleases/070403-01.html> (accessed November 12, 2008).

——. 2008a. "Denso Vision 2015: Achieving Our Goal." *Denso Global*. Kariya, Japan: Denso Corporation. Online available: <http://www.globaldenso.com/en/aboutdenso/vision/> (accessed November 12, 2008).

——. 2008b. "Denso Japan: Capabilities." *About Denso*. Kariya, Japan: Denso Corporation. Online available: <http://www.denso.co.jp/en/aboutdenso/capabilities/> (accessed November 2, 2008).

——. 2008c. "Global Network – Asia & Oceania – Thailand." *Global Denso*. Kariya, Japan: Denso Corporation. Online available: <http://www.denso.co.jp/en/aboutdenso/globalnetwork/asia_oceania/index.html#160> (accessed November 12, 2008).

——. 2008d. "Denso Annual Report 2008, for the Year Ended March 31, 2008" in *Global Denso*. Kariya, Japan: Denso Corporation. Online available: <http://www.globaldenso.com/en/investors/annual/pdf/2008_annual_report.pdf> (accessed November 12, 2008).

Denso-T, Denso Thailand. 2008. "Thai Denso Group." Kariya, Japan: Denso Corporation. Online available: <http://www.denso.co.th/denso/en/about_4.jsp?lang=en> (accessed November 12, 2008).

Dent, Christopher M. 2006. *New Free Trade Agreements in the Asia–Pacific*. New York: Palgrave Macmillan.

Dicken, Peter. 2003. *Global Shift: Reshaping the Global Economic Map in the 21st Century*. New York: Guilford Press.

Doner, Richard F. 1991. *Driving a Bargain: Automobile Industrialization and Japanese Firms in Southeast Asia*. Berkeley, CA: University of California Press.

——. 1997. "Japan in East Asia: Institutions and Regional Leadership." See pp. 197–233 in *Network Power: Japan and Asia*, edited by Peter J. Katzenstein and Takashi Shiraishi. Ithaca: Cornell University Press.

Doner, Richard F., Gregory W. Noble, and John Ravenhill. 2004. "Production Networks in East Asia's Auto Parts Industry." See pp. 159–208 in *Global Production Networking and Technological Change in East Asia*, edited by Shahid Yusuf, M. Anjum Altaf and Kaoru Nabeshima. Washington, DC: The World Bank and Oxford University Press.

Dore, Ronald. 1983. "Goodwill and the Spirit of Market Capitalism." *The British Journal of Sociology* 34, 4: 459–82.

Doz, Yves L., José Santos, and Peter J. Williamson. 2001. *From Global to Metanational: How Companies Win in the Knowledge Economy*. Boston, MA: Harvard Business School Press.

DTC, Danish Trade Council. 2006. "Sector Overview: The Automotive Industry in Thailand, 20 June 2006." Bangkok: Royal Danish Embassy. Online available: <http://www.ambbangkok.um.dk/en/menu/CommercialServices/MarketOpportunities/Sector+overviews/> (accessed May 12, 2007).

——. 2007. "Sector Overview: Sourcing from Thailand, the Electronics Sector." Bangkok: Royal Danish Embassy. Online available: <http://www.ambbangkok.um.dk/NR/rdonlyres/E67766BA-5F8D-4C67-B659-CAEA10B7807/0/SectorOverviewSourcingTheelectronicsindustryapril2007.pdf> (accessed May 12, 2008).

Dunning, John H. 2000. "The Eclectic Paradigm of International Production: A Personal Perspective." See pp. 119–39 in *The Nature of the Transnational Firm*, edited by Christos N. Pitelis and Roger Sugden. New York: Routledge.

Economist. 1997. "Matsushita's China Burn (Interview with Yukio Shotoku)." *Economist* Vol. 344, Issue 8035 (September 30): 75–7.

ED Research. 2007. *FPD Sangyō Gyōkai Chizu: FPD Kanren Kigyō Teikei Senryaku (A Map of the Industrial Circle of Flat Panel Displays: Strategies of Cooperation in Flat Panel Display Firms)*. Tokyo: ED Research.

EEI. 2007. "Thailand's Electrical and Electronics Institute (EEI)." Bangkok: Thailand, Electrical and Electronics Institute (EEI).

Eichengreen, Barry. 2002. "Capitalizing on Globalization." *Asian Development Review* 19, 1: 14–66.

Eichengreen, Barry, and Tong Hui. 2006. "How China Is Reorganizing the World Economy." *Asian Economic Policy Review (AEPR)* 1: 73–97.

EPA. 2003. "List of Japanese Participants in the December 2003 JTEPA Task Force Meeting." Tokyo: Economic Partnership Agreements Section, Ministry of Economy, Trade and Industry (METI). Online available: <http://www.meti.go.jp/policy/trade_policy/epa/data/thailand_report2_japan_e.pdf> (accessed November 13, 2005).

ERIA. 2008. "ERIA Holds Inaugural Governing Board Meeting at ASEAN Secretariat, 3 June 2008." *Documents on Economic Integration*, edited by ASEAN. Jakarta: ASEAN Secretariat. Online available: <http://www.aseansec.org/21606.htm> (accessed November 12, 2008).

Ernst, Dieter. 2000. "Evolutionary Aspects: The Asian Production Networks of Japanese Electronics Firms." See pp. 80–111 in *International Production Networks in Asia: Rivalry or Riches?*, edited by Michael Borrus, Dieter Ernst and Stephan Haggard. New York: Routledge.

——. 2006a. "Searching for a New Role in East Asian Regionalization: Japanese Production Networks in the Electronics Industry." See pp. 161–87 in *Beyond Japan: The Dynamics of East Asian Regionalism*, edited by Peter J. Katzenstein and Takashi Shiraishi. Ithaca, NY: Cornell University Press.

——. 2006b. "Innovation Offshoring: Asia's Emerging Role in Global Innovation Networks." In *East–West Center Special Reports* No. 10, edited by East–West Center. Honolulu, HI: East–West Center. Online available: <http://www.eastwestcenter.org/stored/pdfs/SR010.pdf> (accessed May 6, 2008).

ESCAP. 2007. *Economic and Social Survey of Asia and the Pacific 2007: Surging Ahead in Uncertain Times*. Bangkok: United Nations, Economic and Social Committee for Asia and the Pacific (ESCAP).

ETNA. 2007. "Thailand Exports Half Its Auto Production." *ETNA English News* August 9, 2007. Bangkok: ETNA MCOT News. Online available: <http://etna.mcot.net/query.php?nid=30964> (accessed November 14, 2008).

Fan, Peilei. 2006. "Catching Up Through Developing Innovation Capability: Evidence from China's Telecom-Equipment Industry." *Technovation* 26, 3 (March): 359–68.

Felker, Greg B. 2004. "Global Production and Southeast Asia's Industrialization." See pp. 82–105 in *Asian Regional Governance: Crisis and Change*, edited by Kanishka Jayasuriya. New York: RoutledgeCurzon.

Fourin. 2007. *Nihon Jidōsha Sangyō 2007 [Japan Auto Industry 2007]*. Nagoya, Japan: Fourin.

——. 2008. *Chūgoku Jidōsha shijō no Seihin Kyōsōryoku Hikaku [Comparative Competitive Strength of Products in China's Automotive Market]*. Nagoya, Japan: Fourin.

Frankel, Jeffrey A. 1993. "Is Japan Creating a Yen Bloc in East Asia and the Pacific." See pp. 53–88 in *Regionalism and Rivalry: Japan and the United States in Pacific Asia*, edited by Jeffrey A. Frankel and Miles Kahler. Chicago: University of Chicago Press.

Fruin, Mark. 1998. "Analyzing Pacific Rim Networks and Markets: An Introduction." See pp. 3–31 in *Networks, Markets, and the Pacific Rim*, edited by W. Mark Fruin. Oxford: Oxford University Press.

FTI, Federation of Thai Industries. 2008. "The Textile Industry Club." Bangkok: Federation of Thai Industries (FTI). Online available: <http://www.fti.or.th/Fti%20Project/GroupCallEng.aspx> (accessed November 14, 2008).

Fukao, Kyoji, Keiko Ito, Shigesaburo Kabe, Deqiang Liu, and Fumihide Takeuchi. 2006. "Do Japanese Firms Fail to Catch Up in Localization? An Empirical Analysis Based on Affiliate-level Data of Japanese Firms and a Case Study of Automobile Industry in China." *JCER Discussion Paper Series* No. 99. Tokyo: Japan Center for Economic Research (JCER). Online available: <http://www.jcer.or.jp/eng/pdf/discussion99.pdf> (accessed May 5, 2007).

Fung, K. C., and Hitomi Iizaka. 2004. "Japanese Trade and Direct Investment Links in China and Korea." See pp. 75–114 in *Enhancing Investment Cooperation in Northeast Asia*, edited by Joon-kyung Kim and Chang-jae Lee. Seoul: Korean Institute for International Economic Policy (KIEP).

Gao, Bingzhong. 2000. "Issues on the Legitimacy of Social Organizations." *Social Sciences in China* No. 2. Beijing: Institute of Sociology and Anthropology, Peking University. Online available: <http://base.china-europa-forum.net/rsc/docs/doc_394.pdf> (accessed September 11, 2006).

Gaulier, Guillaume, Françoise Lemoine, and Deniz Ünal-Kesenci. 2004. "China's Integration in Asian Production Networks and Its Implications." In *RIETI Discussion Paper Series* 04-E-033 edited by Trade Research Institute of Economy and Industry (RIETI). Tokyo: RIETI (Research Institute of Economy, Trade and Industry).

Gereffi, Gary. 1995. "Global Production Systems and Third World Development." See pp. 100–42 in *Global Change, Regional Response: The New International Context of Development*, edited by Barbara Stallings. Cambridge: Cambridge University Press.

——. 1999. "International Trade and Industrial Upgrading in the Apparel Commodity Chain." *Journal of International Economics* 48: 37–70.

——. 2001. "Shifting Governance Structures in Global Commodity Chains, with Special Reference to the Internet." *American Behavioral Scientist* 44, 10 (June): 1616–37.

Gereffi, Gary, and Miguel Korzeniewicz (eds). 1994. *Commodity Chains and Global Development*. Westport, CI: Praeger.

Giddens, Anthony. 1990. *The Consequences of Modernity*. Stanford, CA: Stanford University Press.

GM, General Motors. 2008. "Corporate Information, Global Operations: Korea." Detroit, MI: General Motors. Online available: <http://www.gm.com/corporate/about/global_operations/asia_pacific/kore.jsp> (accessed November 12, 2008).

GMB. 2006. *The Handbook of Country Risk 2006: A Guide to International Business and Trade*. London: Global Market Briefings (GMB).

Granovetter, Mark. 1994. "Business Groups." See pp. 453–75 in *The Handbook of Economic Sociology*, edited by Neil Smelser and Richard Swedberg. Princeton: Princeton University Press.

——. 1995. "The Economic Sociology of Firms and Entrepreneurs." See pp. 128–65 in *The Economic Sociology of Immigration: Essays on Networks, Ethnicity, and Entrepreneurship*, edited by Alejandro Portes. New York: Russell Sage.

GTIS, Global Trade Information Service. 2007. *World Trade Analyzer*. Toronto: Global Trade Information Service.

Guyton, L. E. 1996. "Japanese Investments and Technology Transfer to Malaysia." See pp. 70–101 in *Capital, the State, and Late Industrialization*, edited by John Borrego, A. A. Bejar and J. K. S. Jomo. Boulder, CO: Westview Press.

Haggard, Stephen, Sylvia Maxfield, and Ben Ross Schneider. 1997. "Theories of Business and Business–State Relations." See pp. 36–60 in *Business and the State in Developing Countries*, edited by Sylvia Maxfield and Ben Ross Schneider. Ithaca, NY: Cornell University Press.

Hall, Peter A., and David Soskice. 2001. "An Introduction to Varieties of Capitalism." See pp. 1–66 in *Varieties of Capitalism: The Institutional Foundations of Comparative Advantage*, edited by Peter A. Hall and David Soskice. Oxford: Oxford University Press.

Han, Man-Hee. 1994. *Japanese Multinationals in the Changing Context of Regional Policy*. Aldershot, UK: Avebury.

Harvey, David. 1989. *The Condition of Postmodernity: An Enquiry into the Origins of Cultural Change*. Oxford: Blackwell.

Hatch, Walter. 2003. "Japanese Production Networks in Asia: Extending the Status Quo." See pp. 22–56 in *Crisis and Innovation in Asian Technology*, edited by William W. Keller and Richard J. Samuels. Cambridge: Cambridge University Press.

Hatch, Walter, and Kozo Yamamura. 1996. *Asia in Japan's Embrace: Building a Regional Production Alliance*. Cambridge: Cambridge University Press.

Hayashi, Shigeko. 2006. *Japan and East Asian Monetary Regionalism: Towards a Proactive Leadership Role?* New York: Routledge.

Hayashi, Yasuo (JETRO CEO). 2007. "Sustainable Development of Japan–China Economic Relations and Prospects for Business Alliance." JETRO – Home – Outline of JETRO – Speeches by Chairman. Tokyo: Japan External Trade Organization (JETRO). Online available: <http://www.jetro.go.jp/en/jetro/profile/speeches/2007/jun07.pdf> (accessed April 4, 2008).

Henderson, Jeffrey. 2004. "Globalization on the Ground: Global Production Networks, Competition, Regulation and Economic Development." *Working Paper Series* Paper No. 38, Centre on Regulation and Competition, Institute for Development Policy and Management. Manchester: University of Manchester.

Herbig, Paul A. 1995. *Innovation Japanese Style: A Cultural and Historical Perspective*. Westport, CT: Quorum Books.

Hershberg, Eric, Kaoru Nabeshima, and Shahid Yusuf. 2007. "Opening the Ivory Tower to Business: University–Industry Linkages and the Development of Knowledge-intensive Clusters in Asian Cities." *World Development* 35, 6 (June): 931–40.

Hobday, Michael. 1995. *Innovation in East Asia: The Challenge to Japan*. Cheltenham, UK: Edward Elgar.

Hollingsworth, J. Rogers, and Robert Boyer. 1997a. "Coordination of Economic Actors and Social Systems of Production." See pp. 1–47 in *Contemporary Capitalism – The Embeddedness of Institutions*, edited by J. Rogers Hollingsworth and Robert Boyer. Cambridge: Cambridge University Press.

——. 1997b. "From National Embeddedness to Spatial and Institutional Nestedness." See pp. 433–84 in *Contemporary Capitalism: The Embeddedness of Institutions*, edited by J. Rogers Hollingsworth and Robert Boyer. Cambridge: Cambridge University Press.

Hollingsworth, J. Rogers, Philippe G. Schmitter, and Wolfgang Streeck. 1994. "Capitalism, Sectors, Institutions, and Performance." See pp. 3–16 in *Governing Capitalist Economies: Performance and Control of Economic Sectors*, edited by J. Rogers Hollingsworth, Philippe G. Schmitter, and Wolfgang Streeck. New York: Oxford University Press.

Holzhausen, Arne. 2000. "Restructuring à la Japonaise: Supplier Relationships in the Japanese Car Industry in the 1990s." See pp. 51–65 in *Can Japan Globalize? Studies on Japan's Changing Political Economy and the Process of Globalization in Honor of Sung-Jo Park*, edited by Arne Holzhausen. Frankfurt: Physica-Verlag.

Hopkins, T. K., and Immanuel Wallerstein. 1994. "Commodity Chains: Construct and Research." See pp. 17–20 in *Commodity Chains and Global Capitalism*, edited by Gary Gereffi and Miguel Korzeniewicz. Westport, CN: Greenwood Press.

Hosokawa, Daisuke. 2007. "Japan's Business Strategy in Asia after the Boom of Investing in China." *ISBR Small Business Monograph* 16 (January): 1–31.

Humphrey, John, and Hubert Schmitz. 2002. "Developing Country Firms in the World Economy: Governance and Upgrading in Global Value Chains." *INEF Report* (Institute for Development and Peace, University of Duisburg) No. 61: 18–35.

——. 2004. "Governance in Global Value Chains." See pp. 93–109 in *Local Enterprises in the Global Economy: Issues of Governance and Upgrading*, edited by Hubert Schmitz. Northampton, MA: Edward Elgar.

Hurrell, Andrew. 1995. "Regionalism in Theoretical Perspective." See pp. 37–73 in *Regionalism in World Politics: Regional Organization and International Order*, edited by Louise Fawcett and Andrew Hurrell. Oxford: Oxford University.

Imai, Kenichi. 1992. "Japan's Corporate Networks." See pp. 198–230 in *The Political Economy of Japan, Vol. 3: Cultural and Social Dynamics*, edited by Shumpei Kumon and Henry Rosovsky. Stanford, CA: Stanford University Press.

IMF. 2007. *World Economic Outlook Database, October 2007 Edition: Report for Selected Countries and Subjects*. Washington, DC: International Monetary Fund. Online available: <http://www.imf.org/external/pubs/ft/weo/2007/02/weodata/index.aspx> (accessed November 12, 2008).

Inoguchi, Takashi. 2006. "Has the Japanese Model Ceased to Be a Magnet in Asia?" See pp. 204–22 in *Governance and Regionalism in East and Southeast Asia: A Comparative Study*, edited by Ian Marsh. New York: Routledge.

INT, Interview. 1999a. "Interview with Deputy Director, Policy Research Department, Ministry of Foreign Trade and Economic Cooperation (MOFTEC) June 4, 1999." Beijing: Ministry of Foreign Trade and Economic Cooperation (MOFTEC).

——. 1999b. "Interview with Director, Denso (Thailand), April 20, 1999." Samut Prakan, Thailand: Denso Thailand.

——. 1999c. "Interview with Deputy General Manager, Tianjin Denso Engine Electric Products Co. Ltd., June 1, 1999." Tianjin, China: Tianjin Denso Engine Electric Products Co. Ltd.

——. 1999d. "Interview with Isao Masaoka, Deputy Managing Director, Denso Thailand Co. Ltd, July 1, 1999." Samut Prakan, Thailand: Denso Thailand Co. Ltd.

——. 1999e. "Interview with Manager, Engineering Division, Toyota Motor Technical Center TTC, Tianjin Hi-Tech Industry Park, Tianjin, May 31, 1999." Tianjin, China: Toyota Motor Technical Center.

——. 1999f. "Interview with Beijing Representative, Toyota Motor Corporation China Office, June 3, 1999." Beijing: Toyota Motor Corporation Beijing Office.

——. 1999g. "Interview with Director, Japan Chemical Fibers Association (JCFA) March 19, 1999." Tokyo: Japan Chemical Fibers Association (JCFA).

——. 1999h. "Interview with Chief Representative, Toray Korea, Senior Managing Director, Saehan Industries Inc., June 16, 1999." Seoul: Saehan Industries.

——. 1999i. "Interview with General Manager for China, Itochu Corporation, Beijing Liaison Office, January 2, 1999." Beijing: Itochu Corporation, Beijing Liaison Office.

——. 1999j. "Interview with Director, China Chemical Fibers Association (CCFA), June 8, 1999." Beijing: China Chemical Fibers Association (CCFA).

——. 2002a. "Interview with Manager in Chief, International Division, The Japan Chamber of Commerce and Industry, The Tokyo Chamber of Commerce and Industry, November 29, 2002." Tokyo: The Japan Chamber of Commerce and Industry, The Tokyo Chamber of Commerce and Industry.

——. 2002b. "Manager, Government and Industrial Affairs, Tokyo Office, Denso Corporation, December 2, 2002." Tokyo: Denso Corporation.

——. 2002c. "Interview with Manager of China Operations, and Manager of Sales and Marketing Dept 2, Sales Center 3, Display and New Business Planning, Denso Corporation, December 18, 2002." Kariya, Japan: Denso Corporation.

——. 2003. "Interview with Department Head, Qingdao Association of Enterprises with Foreign Investment, Branch of Korean Foreign Invested Enterprises, Qingdao International Bank Office, July 17, 2003." Qingdao, PRC: Korea Trade-Investment Promotion Agency (KOTRA).

——. 2004. "Interview with Vice-President, JETRO Shanghai, June 23, 2004." Shanghai: Japan External Trade Organization (JETRO).

——. 2007a. "Interview with Director General, Department for International Affairs, Development Bank of Japan, January 29, 2007." Tokyo: Development Bank of Japan (DBJ).

——. 2007b. "Interview with Senior Advisor to the Board, Toyota Motor Corp., and Co-Chair, Trade and Investment Liberalization and Facilitation Working Group, APEC Business Advisory Council, February 19, 2007." Tokyo: Toyota Motor Corporation.

——. 2007c. "Interview with Special Advisor, Nikko Citigroup Ltd., Co-chair, ABAC Finance Committee, Asia–Pacific Economic Cooperation (APEC), February 19, 2007." Tokyo: Nikko Citigroup Ltd.

——. 2007d. "Interview with Director, Economic Information Department, JETRO Beijing, January 10, 2007." Beijing: Japan External Trade Organization (JETRO).

Intarakumnerd, Patarapong. 2006. "Thailand's National Innovation System in Transition." See pp. 100–22 in *Asia's Innovation Systems in Transition*, edited by Bengt-Ake Lundvall, Patarapong Intarakumnerd and Jan Vang-Lauridsen. Northampton, MA: Edward Elgar.

IRI, Industrial Research Institute Co. 2005. *ASEAN Jidōsha Sangyō 2005 [ASEAN Auto Industry, 2005]*. Tokyo: Industrial Research Institute [Kōgyō Chōsa Kenkyūjo] Co. Ltd., and NNA (Thailand) Co., Ltd.

Ishii, Ginjiro. 2005. "Current Status and Challenges of Japanese Textiles and Clothing Industry, and Cooperation among Japan, Korea, and China." See pp. 1–14 in *Papers of the 4th Korea, China and Japan Business Forum, October 13–14, 2005*, edited by Federation of Korean Industries (FKI). Seoul: Federation of Korean Industries (FKI).

ITGLWF. 2008. "List of ITGLWF Affiliates and Associates." Singapore: International Textile, Garment, and Leather Workers' Federation (ITGLWF). Online available: <http://www.itglwf.org/pdf/06-09-26%20-%20ITGLWF%20Affiliates%20and%20Associate.pdf> (accessed November 12, 2008).

Ito, Banri, Eiichi Tomiura, and Ryuhei Wakasugi 2007. "Dissecting Offshore Outsourcing and R&D: A Survey of Japanese Manufacturing Firms." *RIETI Discussion Paper Series* 07-E-060. Tokyo: Research Institute of Economy, Trade and Industry (RIETI). Online available: <http://www.rieti.go.jp/jp/publications/dp/07e060.pdf> (accessed November 12, 2008).

Itochu. 2007a. "Itochu's Comprehensive Strength – Outline of Seven Division Companies – Textile Division – Asia." Tokyo: Itochu Corporation. Online available: <http://www.itochu.co.jp/main/div/tex/tex_asia_e.html> (accessed May 12, 2008).

——. 2007b. "News Release 19 March 2007: Startup of Facilities for Production of High-purity Terephthalic Acid in the Daxie Development Zone of Ningbo, China." Tokyo: Itochu Corporation. Online available: <http://www.itochu.co.jp/main/news/2007/news_070319e.html> (accessed November 8, 2008).

——. 2007c. "Itochu Corporation, Investor Information – Annual Financial Statements for the Years Ending March 31, 2007 and 2006." Tokyo: Itochu Corporation. Online available: <http://www.itochu.co.jp/main/ir/financial/2007term/07_ended_e.html> (accessed November 8, 2008).

——. 2008a. "About Itochu China – Future Business Development in China – Division of Textiles." Tokyo: Itochu Corporation. Online available: <http://www.itochu.com.cn/en/b03_en.htm> (accessed November 8, 2008).

——. 2008b. "About Itochu China – Message from President Nobuo Kuwayama." Tokyo: Itochu Corporation. Online available: <http://www.itochu.com.cn/en/index_en.htm> (accessed November 8, 2008).

——. 2008c. "About Itochu China." Tokyo: Itochu Corporation. Online available: <http://www.itochu.com.cn/en/b03_en.htm> (accessed November 8, 2008).

——. 2008d. "Denso Annual Report 2008, for the Year Ended March 31, 2008." In *Global Denso*. Kariya, Japan: Denso Corporation. Online available: <http://www.globaldenso.com/en/investors/annual/pdf/2008_annual_report.pdf> (accessed November 12, 2008).

JAICA, and Akira Inoue. 1989. "A Study on Industrial Sector Development in the Kingdom of Thailand – Second Year Final Draft Report." Edited by JAICA (Japan International Cooperation Agency) Development Survey Project Group. Tokyo: Japan External Trade Organization (JETRO).

JAMA. 2007a. *The Motor Industry of Japan 2007 (Nihon no Jidōsha Kōgyō)*. Tokyo: Japan Automobile Manufacturers Association, Inc. (JAMA).

——. 2007b. *Hand-in-Hand: Partnership in the Auto Industry between ASEAN and Japan, 2007*. Tokyo: Japan Automobile Manufacturers Association, Inc. (JAMA).

——. 2007c. "Nurturing the Future Generation for the Thai Automotive Industry: The Thai-Nichi Institute of Technology." *News from JAMA (Japan Automobile Manufacturers Association), Asia–Pacific Edition* 28, 1 (October): 1–3.

JAPIT. 2006. *Ni-Chū Kankei Kigyō Data 2006 Nenpan [2006 Data on Japanese Enterprise in China]*. Tokyo: Nihon Kokusai Bōeki Zokushin Kyōkai [The Association for the Promotion of International Trade, Japan].

——. 2007. *Ni-Chū Kankei Kigyō Data [2007 Data on Japan–China Enterprise]*. Tokyo: Japan Association for Promotion of International Trade (JAPIT).

JBIC. 2006a. *Survey Report on Overseas Business Operations by Japanese Manufacturing Companies – Results of JBIC FY 2006 Survey: Outlook for Japanese Foreign Direct Investment (18th Annual Survey)*. Tokyo: Japan Bank for International Cooperation (JBIC).

——. 2006b. *Survey Report on Overseas Business Operations by Japanese Manufacturing Companies: Summary – Outlook for Japanese Foreign Direct Investment (18th Annual Survey)*. Tokyo: Japan Bank for International Cooperation (JBIC).

——. 2007. Survey Report on Overseas Business Operations by Japanese Manufacturing Companies – Results of JBIC Survey FY 2007 Survey: Outlook for Japanese Foreign Direct Investment (19th Annual Survey). Tokyo: Japan Bank for International Cooperation (JBIC). Online available: <http://www.jbic.go.jp/autocontents/english/news/2007/000150/full%20document.pdf> (accessed April 4, 2008).

JCCI-B, Japanese Chamber of Commerce and Industry, Bangkok. 2007a. List of Members. Bangkok: Japanese Chamber of Commerce and Industry (JCCI), Bangkok. Online available: <http://www.jcc.or.th/English/Data/yakuin.html> (accessed November 2, 2008).

——. 2007b. "Tai-Nihon Kōgyō Daigaku [Thai-Nichi Institute of Technology, TNI]." *Bangkok JCCI Jōhō* [News of Japanese Chamber of Commerce and Industry, Bangkok] (May): 6–15.

——. 2008. "Name List of Successive Presidents." Edited by JCCI-B. Bangkok: Japanese Chamber of Commerce and Industry (JCCI), Bangkok. Online available: <http://www.jcc.or.th/English/Data/name_list_succ.asp> (accessed November 1, 2008).

JCFA. 2007. *Seni Handobuku 2007 [2007 Textile Handbook]*. Tokyo: Nihon Kagaku Seni Kyōkai (Japan Chemical Fibers Association – JCFA) Shiryō Hanbukai.

JCHEM. 2008. *Kagakuhin Handobuku 2008 [Chemical Fibers Handbook 2008]*. Tokyo: The Heavy and Chemical Industries News Agency (JCHEM).

JEITA. 2007. "Worldwide Electronics and Information Worldwide Electronics and Information Technology Production Trends Technology Production Trends (1st issue)." Tokyo: Japan Electronics and Information Technology Industries Association (JEITA).

JEMA. 2005. *2005 Nenpan JEMA Kaiin Kigyō Kaigai Hōjin Ichiran [2005 List of JEMA Members' Offshore Subsidiaries]*. Tokyo: JEMA (The Japan Electrical Manufacturers' Association; Nihon Denki Kōgyōkai).

——. 2007. "JEMA Vision." Tokyo: Japan Electrical Manufacturers' Association (JEMA).

JETRO. 2004. "JETRO White Paper on International Trade and Foreign Direct Investment 2004 (Summary)." Tokyo: Overseas Research Department, Japan External Trade Organization (JETRO).

——. 2006a. "JETRO White Paper on International Trade and Foreign Direct Investment 2006 (Summary): Japanese Corporate Activity in New Growth Markets and the Emerging East Asian Free Trade Zone." Tokyo: Overseas Research Department, Japan External Trade Organization (JETRO).

——. 2006b. Japanese Trade in 2005. Tokyo: Japan External Trade Organization (JETRO). Online available: <http://www.jetro.go.jp/jpn/stats/data/pdf/trade2005.pdf> (accessed April 4, 2007).

——. 2007a. "White Paper on International Trade and Foreign Direct Investment 2007: Increasing Utilization of Asian FTAs and Growth Strategies for Japanese Companies." Tokyo: Japan External Trade Organization (JETRO). Online available: <http://www.jetro.go.jp/en/stats/white_paper/2007.pdf> (accessed November 2, 2008).

——. 2007b. FY 2006 Survey of Japanese Firms' International Operations. Tokyo: Japan External Trade Organization (JETRO).

——. 2007c. *Japan's Total Outward FDI by Country/Region (Balance of Payments Basis, Net and Flow)*. Tokyo: JETRO (Japan External Trade Organization). Online available: <http://www.jetro.go.jp/en/stats/statistics/bpfdi_01_e.xls> (accessed November 4, 2008).

——. 2008. *Japan's Total Outward FDI by Country/Region (International Investment Position) FDI Outward Stock*. Tokyo: JETRO (Japan External Trade Organization). Online available: <http://www.jetro.go.jp/en/stats/statistics/iipfdi_01_e.xls>_(accessed November 2, 2008).

JICA. 2003a. "Country Study for Japan's Official Development Assistance to the Kingdom of Thailand: From 'Development Assistance' to a 'New Cooperation.'" Tokyo: Institute for International Cooperation, Japan International Cooperation Agency (JICA).

——. 2003b. "From 'Development Assistance' to a 'New Cooperation' – Country Study for Japan's Official Development." Tokyo: Institute for International Cooperation Assistance to the Kingdom of Thailand, Japan International Cooperation Agency (JICA). Online available: <http://www.jica.go.jp/english/publication/studyreport/country/thailand/pdf/thai_02.pdf> (accessed April 4, 2006).

Jirapet, Krik-krai, and Akira Amari. 2007a. "Attachment 3: Automotive Human Resources Development Institute Project (AHRDIP)." See pp. 5–8 in Joint Statement of the Minister of Commerce Kingdom of Thailand and the Minister of Economy, Trade and Industry of Japan, April 3, edited by METI Minister (Japan) Minister of Commerce (Thailand). Bangkok: Ministry for Foreign Affairs (MFA), Kingdom of Thailand. Online available: <http://www.mfa.go.th/jtepa/asset/joint_statement_moc-mati_eng.pdf> (accessed November 12, 2008).

——. 2007b. "Attachment 7: Textile and Apparel Cooperation under Japan–Thailand Economic Cooperation Agreement." See pp. 12–13 in Joint Statement of the Minister of Commerce Kingdom of Thailand and the Minister of Economy, Trade and Industry of Japan, April 3, edited by METI Minister (Japan) Minister of Commerce (Thailand). Bangkok: Ministry for Foreign Affairs (MFA), Kingdom of Thailand. Online available: <http://www.mfa.go.th/jtepa/asset/joint_statement_moc-mati_eng.pdf> (accessed November 12, 2008).

Joint Study Group, China, Japan and Korea. 2004. The Report of the Joint Study Group on the Possible Trilateral Investment Arrangements among China, Japan and Korea. Tokyo: Japan, Ministry of Economy, Trade and Industry (METI).

Jones, Randall, and Tadashi Yokoyama. 2006. "Upgrading Japan's Innovation System to Sustain Economic Growth." *OECD Working Papers* No. 527. Paris: OECD Economics Department, Organization for Economic Co-operation and Development (OECD).

JSBRI. 2006. "2006 White Paper on Small and Medium Size Enterprises in Japan: Small and Medium Size Enterprises at a Turning Point – Strengthening Ties with Overseas Economies and Population Decline in Japan." Tokyo: Japan Small Business Research Institute (JSBRI), Small and Medium Enterprise Agency, Japanese Ministry of Economy, Trade and Industry (METI).

——. 2007. "2007 White Paper on Small and Medium Enterprises in Japan Small and Medium Enterprises: Harnessing Regional Strengths and Confronting the Changes." Tokyo: Japan Small Business Research Institute (JSBRI), Small and Medium Enterprise Agency, Ministry of Economy, Trade and Industry (METI). Online available: <http://www.chusho.meti.go.jp/pamflet/hakusyo/h19/download/2007hakusho_eng.pdf> (accessed April 4, 2008).

JTECS. 2007. "JTECS Homepage: JTECS and Its Activities." Bangkok: Japan–Thailand Economic Cooperation Society (JTECS). Online available: <http://www.jtecs.or.jp/e/jtecs/html> (accessed November 7, 2008).

JTEPO. 2007. *Implementation Agreement of the Japan–Thailand Economic Partnership Agreement*. Bangkok: Japan–Thailand Economic Partnership Office (JTEPO) Ministry of Foreign Affairs, Thai Ministry of Foreign Affairs (MFA). Online available: <http://www.mfa.go.th/jtepa/asset/jtepa_implementing_agreement_eng.pdf> (accessed November 8, 2008).

Kameyama, Yoshihiro. 2007. "Chapter 5: Korea." *EAEP, East Asian Economic Perspectives* 18, 1 (February): 51–62.

KCFA. 2006. "ACFIF Vision Summary Results of the 6th Conference of the Asian Chemical Fiber Industries Federation in Seoul, Korea, May 18–19, 2006." Seoul: Korea Chemical Fibers Association (KCFA).

——. 2008. *Chemical Fibers Handbook 2008*. Seoul: Korea Chemical Fibers Association (KCFA). Online available: <http://www.kcfa.or.kr/main/index.php> (accessed November 8, 2008).

Keohane, Robert O., and Joseph S. Nye. 2000. "Introduction: Governance in a Globalizing World." See pp. 1–44 in *Governance in a Globalizing World*, edited by Joseph S. Nye and John D. Donahue. Washington: Brookings Institution Press.

Kikkawa, Takeo. 2000. "Strategies of Japanese Corporations Facing Global Competition: The Cases of Sony and Toray." *Joint Research Project – Discussion Paper* No. 6. Tokyo: Social Science Institute, University of Tokyo. Online available: <http://project.iss.u-tokyo.ac.jp/kikkawa/iss-6.pdf> (accessed April 4, 2008).

Kim Dohyung. 2007. "On the Patterns of Intra-industry Trade and Industrial Cooperation with a Korea–Japan FTA." *ERINA Report* Vol. 76 (July): 7–25.

Kim, Doo-Jin, and Young-Chan Kim. 2006. *Newly Industrializing Economies and International Competitiveness: Market Power and Korean Electronics Multinationals*. New York: Palgrave Macmillan.

Kim, Linsu. 1997. *Imitation to Innovation: The Dynamics of Korea's Technological Learning*. Boston, MA: Harvard Business School Press.

Kim, Samuel S. 2006. "Inter-Korean Relations in Northeast Asian Geopolitics." See pp. 167–82 in *Korea at the Center: Dynamics of Regionalism in Northeast Asia*, edited by Charles K. Armstrong, Gilbert Rozman, Samuel S. Kim and Stephen Kotkin. New York: M. E. Sharpe.

Kline, Stephen J., and Nathan Rosenberg. 1986. "An Overview of Innovation." See pp. 275–306 in *The Positive Sum Strategy: Harnessing Technology for Economic Growth*, edited by Ralph Landau and Nathan Rosenberg. Washington, DC: National Academy Press.

KNSO. 2006. *International Statistics Yearbook 2006 (Gugje Tonggye Yeongam 2006)*. Daejeon: Korea National Statistics Office (KNSO, Tonggyecheong).

——. 2008. *International Statistics, Production of Electronic Merchandise – Korea Statistical Information Service*. Deajeon: Korea National Statistics Office (KNSO, Tonggyecheong). Online available: <http://www.kosis.kr> (accessed November 14, 2008).

Kodama, Fumio, and Jun Suzuki. 2007. "How Japanese Companies Have Used Scientific Advances to Restructure Their Businesses: The Receiver-active National System of Innovation." *World Development* 35, 6 (June): 976–0.

Kodama, Mitsuru. 2005. "Innovation through Dialectical Leadership: Case Studies of Japanese High-tech Companies." *Journal of High Technology Management Research* 16: 137–56.

——. 2007. "Innovation through Boundary Management: A Case Study in Reforms at Matsushita Electric." *Technonovation* 27, 1–2 (January–February): 15–29.

KOFOTI. 2008. "Textile Industry: Industry Review." Seoul: Seoul: Korea Federation of Textile Industries (KOFOTI). Online available: <http://www.kofoti.or.kr/eng/textile/review02.php> (accessed November 4, 2008).

Kohpaiboon, Archanun. 2006. "Foreign Direct Investment and Technology Spillover: A Cross-industry Analysis of Thai Manufacturing." *World Development* 34, 3 (March): 541–6.

Koike, Kazuo. 1997. "The Internationalization of the Japanese Firm: Japanese Working Practices and Indigenous Asian Workplaces." See pp. 44–80 in *The Political Economy of Japanese Society, Vol. 2: Internationalization and Domestic Issues*, edited by Banno Junji. New York: Oxford University Press.

Koike, Yoichi. 1998. "Globalization of Japanese Subcontract Transaction and Asian Suppliers." See pp. 91–115 in *Learning, Liberalization, and Economic Adjustment*, edited by Mitsuhiro Kagami, John Humphery and Michael Piore. Tokyo: Institute of Developing Economies.

Kosaka, Kenji, Minister of MEXT (Minister for Education, Culture, Sports, Science and Technology). 2007. "Toward Japanese Science and Technology that Contributes to People's Prosperity in the 21st Century." *Science & Technology in Japan* 25: 2–7.

KOTRA. 2003. "Investment: Korea Success Story – Toray Saehan Inc." *Invest Korea*

(November–December). Seoul: Korea Trade Promotion Agency (KOTRA). Online available: <http://www.ikjournal.com/InvestKoreaWar/work/journal/content/content_main.jsp?code=4290206> (accessed November 4, 2008).

——. 2004. "They Invested Here – STEMCO Ltd: Interview with CEO Park Kyu-bok." *Invest Korea* (September–October). Seoul: Korea Overseas Trading Association (KOTRA). Online available: <http://www.ikjournal.com/InvestKoreaWar/work/journal/content/content> (accessed November 20, 2008).

——. 2007. "Multinational Auto Parts Manufacturers in Korea." *Invest Korea – Content* (4 October). Seoul: Korea Trade Promotion Organization (KOTRA). Online available: <http://www.ikjournal.com/?code=102020204> (accessed November 4, 2008).

——. 2008a. "Toshiba Samsung Storage Technology Corporation and Hitachi-LG Data Storage, Inc." *Invest Korea: Success Stories* (9 May). Seoul: Korea Trade Promotion Organization (KOTRA).

——. 2008b. "Automobiles and Auto Parts." *Invest Korea*. Updated October 8, 2008. Seoul: Korea Overseas Trade Promotion Organization (KOTRA). Online available: <http://www.investkorea.org/?code=102020201> (accessed November 4, 2008).

——. 2008c. "R&D Support Systems." *Invest Korea – Webpage – Business Opportunities – R&D*. Updated March 10, 2008. Seoul: Korea Trade Promotion Organization (KOTRA). Online available: <http://www.investkorea.org/InvestKoreaWar/work/ik/eng/bo/content_print.jsp?code=102020801> (accessed November 4, 2008).

KT, Park Hyong-ki. 2007. "Kyonggi Signs MOU with Japan Auto Parts Maker." *Korea Times – Business Page*. March 1, 2007. Seoul: Hankook Ilbo Co.

KT. 2008a. "Korean IT Giants Facing 'Survival Game.'" *Korea Times*. March 29, 2008. Seoul: Hankook Ilbo Co. Online available: <http://www.koreatimes.co.kr/www/news/biz/2008/03/123_20375.html> (accessed March 29, 2008).

——. 2008b. "Panasonic to Bolster Tie-up with Samsung in Blu-ray." *Korea Times*. April 20, 2008. Seoul: Hankook Ilbo Co. Online available: <http://www.koreatimes.co.kr/www/news/biz/2008/04/123_22828.html> (accessed April 20, 2008).

KT, Kim Hyun-cheol. 2008c. "Korean Auto Power Underscores Tiny Foreign Presence." *Korea Times – Business Page*. May 14, 2008. Seoul: Hankook Ilbo Company.

Kumar, Krishan. 1995. *From Post-industrial to Post-modern Society: New Theories of the Contemporary World*. Oxford, UK: Blackwell Publishers.

Kumon, Shumpei. 1992. "Japan as a Network Society." See pp. 109–41 in *The Political Economy of Japan, Volume 3: Cultural and Social Dynamics*, edited by Shumpei Kumon and Henry Rosovsky. Stanford, CA: Stanford University Press.

Kuroda, Haruhiko. 2007. "Prosperity with Inclusiveness, Growth with Sustainability: A Shared Aspiration for Development." In *40th Annual Meeting Address to the Board of Governors, Asian Development Bank*. Manila: Asian Development Bank.

Kurokawa, Kiyoshi. 2007. "Essentials of Innovation." *The Japan Journal* (June): 14–15.

Lee I. K. (Inkoo). 2006. "The Evolution of and Prospects for Economic Relations between the ROK and China." *ERINA Report* 72 (November): 10–13.

Lee, Keun, and Chaisung Lim. 2001. "Technological Regimes, Catching-up and Leapfrogging: Findings from the Korean Industries." *Research Policy* 30: 459–83.

Lee Kong-Rae. 2006. "Performance and Sources of Industrial Innovation in Korea's Innovation System." See pp. 178–99 in *Asia's Innovation Systems in Transition*, edited by Bengt-Ake Lundvall, Patarapong Intarakumnerd and Jan Vang-Lauridsen. Northampton, MA: Edward Elgar.

Lemoine, Françoise, and Deniz Ūnal-Kesenci. 2004. "Assembly Trade and Technology Transfer: The Case of China." *World Development* 32, 5: 829–50.

Lincoln, Edward J. 2005. "Economic Regionalism in East Asia." See pp. 60–74 in *Regionalism in Northeast Asia: Opportunities and Challenges*, edited by Hyungdo Ahn and Yong Shik Choo. Seoul: Korea Institute for International Economic Policy (KIEP) and Johns Hopkins SAIS.

Lincoln, James R. 2006. "Interfirm Networks and the Management of Technology and Innovation in Japan." See pp. 214–34 in *Recovering from Success: Innovation and Technology Management in Japan*, edited by D. H. Whittaker and Robert E. Cole. New York: Oxford University Press.

López-Martínez, Roberto, and Andrea Piccaluga. 2001. "Introduction: The Meso-Foundations of National Innovation Systems." See pp. 1–14 in *Knowledge Flows in National Systems of Innovation: A Comparative Analysis of Sociotechnical Constituencies in Europe and Latin America*, edited by Roberto López-Martínez and Andrea Piccaluga. Northampton, MA: Edward Elgar.

Lynn, Leonard H., and Timothy J. McKeown. 1988. *Organizing Business: Trade Associations in America and Japan*. Washington, DC: American Enterprise Institute for Public Policy Research.

McDonald, Frank, and Giovanna Vertova. 2002. "Clusters, Industrial Districts and Competitiveness." See pp. 38–64 in *Global Competition and Local Networks*, edited by Rod B. McNaughton. Burlington, VT: Ashgate.

MacDuffie, John Paul, and Susan Helper. 2006. "Collaboration in Supply Chains, With and Without Trust." See pp. 417–66 in *The Firm as a Collaborative Community: Reconstructing Trust in the Knowledge Economy*, edited by Charles Heckscher and Paul S. Adler. Oxford, UK: Oxford University Press.

McMichael, Philip. 2008. *Development and Social Change – A Global Perspective, Fifth Edition*. Thousand Oaks, CA: Pine Forge Press.

McNamara, Dennis L. 1990. *The Colonial Origins of Korean Enterprise, 1910–1945*. Cambridge: Cambridge University Press.

——. 1995. *Textiles and Industrial Transition in Japan*. Ithaca: Cornell University Press.

——. 1996. "Corporatism and Cooperation among Japanese Labor." *Comparative Politics* Vol. 28, No. 4 (July): 379–98.

——. 1999. "Korean Capitalism." See pp. 139–50 in *Corporatism and Korean Capitalism*, edited by Dennis L. McNamara. New York: Routledge.

——. 2002. *Market and Society in Korea: Interest, Institution and the Textile Industry*. London: Routledge.

——. 2005. "Industry and National Identity: Globalizing the Korean Auto Industry." See pp. 318–28 in *Korean Attitudes Toward the United States*, edited by David Steinberg. New York: M. E. Sharpe.

——. 2006. "New Places but Old Spaces: Knowledge Hierarchies Among Asian SMEs Abroad." *Organization: The Critical Journal of Organization, Theory and Society* 13, 4: 549–67.

Maeda, Hiroshi. 1996a. "Toray no Gurobarizashon to Gijutsu Iten [Globalization and Technology Transfer at Toray]: Part I." *JTECS Thai Kōgyō Jōhō Faiyaru* No. 98 (May): 1–7.

——. 1996b. "Toray no Gurobarizashon to Gijutsu Iten [Globalization and Technology Transfer at Toray]: Part II." *JTECS Thai Kōgyō Jōhō Faiyaru* No. 99 (June): 11–15.

Maeda, Yasuo. 1999. "Textile Industry." See pp. 177–98 in *Japan's Industrial Technology Development: The Role of Cooperative Learning and Institutions*, edited by Yoshitaka Okada. Tokyo: Springer-Verlag.

Manning, Richard. 2007. *2006 Development Co-operation Report*. Paris: OECD

(Organization for Economic Co-operation and Development), Development Assistance Committee.

Markusen, Ann. 1996. "Sticky Places in Slippery Space: A Typology of Industrial Districts." *Economic Geography* 72: 293–313.

MCOT. 2006. "Commerce Minister Eyes Japan as Leading Business Partner." *MCOT Thai News Agency*. January 30. Bangkok: MCOT Thai News Agency. Online available: <http://www.mfa.go.th/jtepa/archives/news_2029.html> (accessed November 4, 2008).

MEI. 2008. *Annual Report for the Year Ending March 31, 2007: On the Road to Excellence*. Osaka: Matsushita Electric Industrial (MEI) Co. Online available: <http://ir-site.panasonic.com/annual/2007/pdf/matsushita_ar2007_e.pdfF:/NTjapan/Matsushita_07_data> (accessed November 4, 2008).

MERI, Mitsubishi Economic Research Institute. 2006a. "East Asian Economic Integration and the Japanese Response." *Survey of Economic Conditions in Japan* 914 (February): 6–7.

——. 2006b. "Direction of Overseas Investment by the Japanese Manufacturing Industry." *Survey of Economic Conditions in Japan* 920 (August): 6–9.

——. 2008. "Research and Development Investments, the Key to Chinese Industry." *MERI's Monthly Circular* (August): 6–8.

METI. 2001. "White Paper on International Economy and Trade, 2001." Tokyo: Japanese Ministry of Economy, Trade and Industry (METI).

——. 2004. "White Paper on International Economy and Trade 2004 (Key Points): Toward a New Value Creation Economy." Tokyo: International Economic Affairs Bureau, Trade Policy Bureau, Japanese Ministry of Economy, Trade and Industry (METI).

——. 2005. "White Paper on International Economy and Trade 2005 (Overview). Toward Sustained Potential for Growth: Enhanced Productivity Through Globalization and Japan as an Investment Powerhouse." Tokyo: International Economic Affairs Bureau, Trade Policy Bureau, Japanese Ministry of Economy, Trade and Industry (METI). Online available: <www.meti.go.jp/english/report/downloadfiles/gWP2005s.pdf> (accessed October 6, 2007).

——. 2006a. New Economic Growth Strategy, Outline (Provisional Translation). Tokyo: Japanese Ministry of Economy, Trade and Industry (METI). Online available: <http://www.meti.go.jp/english/report/downloadfiles/NewEconomicGrowthStrategy(outline).pdf> (accessed April 4, 2007).

——. 2006b. "Global Economic Strategy (Summary)." Tokyo: International Economic Affairs Division, Trade Policy Bureau, Japanese Ministry of Economy, Trade and Industry (METI).

——. 2007a. "Trends in Overseas Subsidiaries: Quarterly Survey of Overseas Subsidiaries, Period from October to December 2007: Actual Result Breakdown Lists for ASEAN 4, NIEs 3, and China (including Hong Kong)." Tokyo: Enterprise Statistics Office, Ministry of Economy, Technology and Industry (METI). Online available: <http://www.meti.go.jp/english/statistics/tyo/genntihou/pdf/h2c3h3ye.pdf> (accessed April 4, 2008).

——. 2007b. "White Paper on International Economy and Trade 2007 (Overview): Japan's Trade Strategy on Improving Industrial Productivity and Accelerating Economic Growth – Dynamic Growth of East Asia and Globalization of Service Industries." Tokyo: International Economic Affairs Bureau, Trade Policy Bureau, Japanese Ministry of Economy, Trade and Industry (METI). Online available: <http://www.jetro.go.jp/en/stats/white_paper/2007.pdf> (accessed June 21, 2008).

——. 2007c. "Summary of the 36th Survey on Overseas Business Activities (conducted in July 2006)." Tokyo: Enterprise Statistics Office, Japanese Ministry of Economy,

Technology and Industry (METI). Online available: <http://www.meti.go.jp/english/statistics/tyo/kaigaizi/pdf/h2c400ie.pdf> (accessed May 27, 2008).

——. 2007d. "White Paper on International Economy and Trade 2007: Japan's Trade Strategy on Improving Industrial Productivity and Accelerating Economic Growth – Dynamic Growth of East Asia and Globalization of Service Industries." Tokyo: International Economic Affairs Bureau, Trade Policy Bureau, Japanese Ministry of Economy, Trade and Industry (METI). Online available: <http://www.meti.go.jp/english/report/data/gWT2007fe.html> (accessed June 21, 2008).

MEW. 2006a. "Annual Report 2006, Matsushita Electric Works, Ltd." Osaka: Matsushita Electric Works (MEW). Online available: <http://www.mew.co.jp/e/corp/ir/annual/2006repo/pdfs/15.pdf> (accessed April 4, 2007).

——. 2006b. "News Release 2006: Joint Entry to Green Building Engineering Business in China – Investing in Beijing Tsinghua Tellhow Intelligence Tech, a Core Member of Tellhow Group." Osaka: Matsushita Electric Works (MEW). Online available: <http://www.mew.co.jp/e/corp/news/2006/0609-01.htm#pagetop> (accessed April 4, 2007).

MEXT. 2006a. "2006 White Paper on Science and Technology: Statistics – Technology Trade, Tables 2.2, 2.3." See pp. 1–22 (Appendix) in "White Paper on Science and Technology 2006: Challenge for Building a Future Society: The Role of Science and Technology in an Aging Society with Fewer Children," edited by MEXT. Tokyo: Japan, Ministry of Education, Culture, Sports, Science and Technology (MEXT).

——. 2006b. "2006 White Paper on Science and Technology: Challenge for Building a Future Society – the Role of Science and Technology in an Aging Society with Fewer Children." Tokyo: Japan, Ministry of Education, Culture, Sports, Science and Technology (MEXT). Online available: <http://www.mext.go.jp/english/news/2007/03/07022214/002/001.htm> (accessed April 4, 2007).

——. 2007. "2007 White Paper on Science and Technology: Results of Promotion of Science and Technology – Creation, Utilization and Succession of Knowledge." Tokyo: Japan, Ministry of Education, Culture, Sports, Science and Technology (MEXT).

MFA, Ministry of Foreign Affairs Kingdom of Thailand. 2007. "Implementation Agreement of the Japan–Thailand Economic Partnership Agreement." Bangkok: Ministry of Foreign Affairs, Kingdom of Thailand (MFA). Online available: <http://www.mfa.go.th/jtepa/asset/jtepa_implementing_agreement_eng.pdf> (accessed April 4, 2008).

Miyagawa, Makio, and Pisan Manawapatand. 2003. "Japan–Thailand Economic Partnership Agreement Task Force Report December, 2003." Bangkok: Ministry of Foreign Affairs.

Mizuno, Junko. 2006. "Nikkei Shinshutsu Kigyō to Europukei Shinshutsu Kigyō no Gijutsu Tenkan to Kokusai Fungyō no Chigai [Differences in Technology Transfer and Division of Labor between Japanese Enterprise Abroad and European Enterprise Abroad]." *Wuurudo Torendo [World Trends]* (October): 36–43.

Mochizuki, Mike M. 1995. "Japan as an Asia–Pacific Power." See pp. 124–59 in *East Asia in Transition: Toward a New Regional Order*, edited by Robert S. Ross. Armonk, NY: M. E. Sharpe.

MOCIE. 2008. *Numbers at a Glance: Key Figures*. Seoul: Seoul: Korea Ministry of Commerce, Industry and Energy (MOCIE).

MOF-Japan. 2007. "Outward/Inward Direct Investment 2006 V. Y. (Heisei 18): Direct Investment Abroad (by Country and Industry)." Tokyo: Ministry of Finance. Online available: <http://www.mof.go.jp/bpdata/ed2bop.csv> (accessed April 4, 2007).

MOFA. 2003. *Japan–Thailand Economic Partnership Agreement Task Force Report December 2003*. Tokyo: Japan Ministry of Foreign Affairs (MOFA). Online available: <http://www.mfa.go.th/jtepa/asset/reportDec11.pdf> (accessed November 4, 2008).

——. 2005. *Japan–Thailand Economic Partnership Agreement Joint Press Statement – September First: Attachment 5 (Cooperation)*. Tokyo: Japan Ministry of Foreign Affairs (MOFA). Online available: <http://www.mofa.go.jp/region/asia-paci/thailand/joint0509/joint0509_5.txt> (accessed November 4, 2008).

——. 2006. "Japan's Official Development Assistance (ODA) White Paper 2006." Tokyo: Japan Ministry of Foreign Affairs, Urban Connections.

Monden, Yasuhiro 1998. *Toyota Production System: An Integrated Approach to Just-In-Time*. Norcross, GA: Engineering and Management Press.

Moon, Chung-in. 1991. "Managing Regional Challenges: Japan, the East Asian NICs and New Patterns of Economic Rivalry." *Pacific Focus* (Fall): 23–47.

——. 2002. "Managing Regional Challenges: Japan, the East Asian NICs, and New Patterns of Economic Rivalry." *Pacific Focus* Vol. 6, No. 2: 23–47.

Morris-Suzuki, Tessa. 1992. "Japanese Foreign Direct Investment and the International Division of Knowledge." See pp. 135–52 in *Japan's Foreign Investment and Asian Economic Interdependence: Production, Trade, and Financial Systems*, edited by Shojiro Tokunaga. Tokyo: University of Tokyo Press.

——. 1994. *The Technological Transformation of Japan: From the Seventeenth to the Twenty-first Century*. Cambridge: Cambridge University Press.

Motohashi, Kazuyuki. 2003. "The Japanese Model: Shifts in Comparative Advantage due to the IT Revolution and Modularization." *Journal of Japanese Trade & Industry* (November–December): 30–5.

——. 2006. "R&D of Multinationals in China: Structure, Motivations and Regional Differences." *RIETI Paper Series* 06-E-005. Tokyo: Research Institute of Economy, Trade and Industry (RIETI), Japanese Ministry of Economy, Trade and Industry (METI).

Motohashi, Kazuyuki, and Xiao Yun. 2005. "China's Innovation System Reform and Growing Industry and Science Linkages." *RIETI Discussion Paper Series* 05-E-011. Tokyo: Research Institute of Economy, Trade and Industry (RIETI), Japanese Ministry of Economy, Trade and Industry (METI).

Mukoyama, Hidehiko. 2005. "The Role of South Korea's Industrial Structure in Its Trade Deficit with Japan: Observations Based on the Skyline Map." *RIM Pacific Business and Industries* 5, 16: 18–40.

Nagashima, Toru. 2007. "Asia: Center of World Textile Output – Technology, Soft Power Propel Teijin Growth – Interview with Nagashima Toru, President and CEO, Teijin Ltd." *Japan Spotlight* (January–February): 32–5.

Nahapiet, J., and S. Ghoshal. 1998. "Social Capital, Intellectual Capital, and the Organizational Advantage." *Academy of Management Review* 23, 2: 242–66.

Nakamura, Takeo, and Toshiaki Shinohara. 2007. "External Aspects of East Asian Economics and Finance in Light of Growing Regional Integration." See pp. 1–19 in *Bank of Japan Research Reports*. Tokyo: Bank of Japan.

Narula, Rajneesh, and John H Dunning. 2000. "Industrial Development, Globalization and Multinational Enterprises: New Realities for Developing Countries." *Oxford Development Studies* 28, 2 (June): 141–67.

NBSC. 2007a. *China Statistical Yearbook 2007 – Main Indicators of State-owned and Non-state-owned Enterprises above Designated Size by Sector (2006)*. Beijing: National Bureau of Statistics (NBSC) China. Online available: <http://www.stats.gov.cn/tjsj/ndsj/2007/indexeh.htm> (accessed October 21, 2008).

——. 2007b. *China Statistical Yearbook 2007 – Main Indicators of Industrial Enterprises with Hong Kong, Macao, Taiwan and Foreign Funds by Industrial Sector (2006)*. Beijing:

National Bureau of Statistics China (NBSC). Online available: <http://www.stats.gov.cn/tjsj/ndsj/2007/indexeh.htm> (accessed October 21, 2008).

NFTECB, Ningbo Foreign Trade and Economic Cooperation Bureau. 2005. "Report December 19, 2005: Youngor Allies with Itochu and Marzotto." Ningbo, Zhejiang, PRC: Ningbo Foreign Trade and Economic Cooperation Bureau (NFTECB). Online available: <http://big.nbfet.gov.cn/ztpd/detail.phtml?path=&catId=&newId=1768&> (accessed October 21, 2008).

Nichū, Jidōsha Sangyō Hatten Kyōryokukai. 1995. "Dai 3kai Nichū Jidōsha Sangyō Hatten Kyōryokukai Hōkokusho [Report of the Third Meeting of the Japan–China Auto Industry Development Association, July 23–5, 1995]." Tokyo: Japanese Ministry of International Trade and Industry (MITI).

Nishimura, Hidetoshi. 2007a. "Asia's New Role in Global Economy and the Relevance of Asian Economic Integration." See pp. 157–62 in *Sixth High-Level Conference on Asian Economic Integration: Agenda for the East Asia Summit*. New Delhi: Office of the Special Assistant to the Chairman on ERIA (Economic Research Institute for ASEAN and East Asia) Matters, Japan External Trade Organization (JETRO).

——. 2007b. "Closing Address." See pp. 77–81 in *Report of ERIA Symposium in Singapore: Opportunities and Challenges towards Regional Integration in East Asia*. Singapore: Office of the Special Assistant to the Chairman on ERIA (Economic Research Institute for ASEAN and East Asia) Matters, Japan External Trade Organization (JETRO).

Niyomsilpa, Sakkarin. 2008. "Industry Globalized: The Automotive Sector." See pp. 17–60 in *Thai Capital after the 1997 Crisis*, edited by Pasuk Phongpaichit and Chris Baker. Chiangmai, Thailand: Silkworm Books.

Nonaka, Ikujiro, Noboru Konno, and Ryoko Toyama. 2001. "Emergence of 'Ba:' A Conceptual Framework for the Continuous and Self-transcending Process of Knowledge Creation." See pp. 13–29 in *Knowledge Emergence*, edited by Ikujiro and Toshiro Nishiguchi Nonaka. Oxford: Oxford University Press.

NSOT. 2007. *2007 Report of the Labor Force Survey*. Bangkok: Ministry of Information and Communication Technology, National Statistical Office, Thailand (NSOT). Online available: <http://web.nso.go.th/eng/en/stat/lfs_e/table2_0107.xls> (accessed October 21, [illegible]

NSTDA. 2008. "Main Information: Software Microchip Electronics." Bangkok: N[illegible] Science and Technology Development Agency (NSTDA). Online available: <[illegible] www.nstda.or.th/en/index.php?option=com_content&task=view&id=267&Itemid=103> (accessed October 21, 2008).

OECD. 2000. *A New Economy? The Changing Role of Innovation and Information Technology in Growth: Information Society*. Paris: Organization for Economic Cooperation and Development.

——. 2006. *Main Science and Technology Indicators 2006–2: Key Figures*. Paris: OECD, Organization of Economic Cooperation and Development.

——. 2007. *OECD Factbook 2007: Economic, Environmental, and Social Statistics*. Paris: Organization of Economic Cooperation and Development (OECD).

OECD/MOST-C. 2007. "OECD Reviews of Innovation Policy: China, Synthesis Report 2007." Paris: Organization for Economic Co-operation and Development, and Ministry of Science and Technology (MOST-C), China. Online available: <http://www.oecd.org/dataoecd/54/20/39177453.pdf> (accessed October 21, 2008).

Ohashi, Hideo. 2006. "The Response of Japanese Business to the Rise of China: Competitiveness of Manufacturing Industries." See pp. 203–39 in *Japanese Responses to Globalization: Politics, Security, Economics and Business*, edited by Glen D. Hook and Harukiyo Hasegawa. Basingstoke: Palgrave Macmillan.

Ohmori, Tatsuya. 1999. "Auto Parts Industry." See pp. 103–32 in *Japan's Industrial Technology Development: The Role of Cooperative Learning and Institutions*, edited by Yoshitaka Okada. Tokyo: Springer-Verlag.

Oizumi, Keiichiro. 2008a. "Thailand Robust Exports Drive Growth in 2007." *JRI Monthly Report* 8, 84 (March): 3–4.

——. 2008b. "Thailand: Robust Exports Drive Growth in 2007." *JRI Monthly Report* Vol. 8 No. 84 (March): 3–4.

Okada, Yoshitaka. 1999. "Cooperative Learning and Technological Development in Japan." See pp. 7–41 in *Japan's Industrial Technology Development: The Role of Cooperative Learning and Institutions*, edited by Yoshitaka Okada. Tokyo: Springer-Verlag.

——. 2001. "Cooperative Learning and Japan's Techno-Governance Structure: Exploratory Case Studies." *Sophia International Review* 23: 19–42.

Osono, Emi. 2003. "Toyota Motor Corporation Asia: A Historical Challenge." See pp. 1–35 in *Hitotsubashi University ICN Case Studies A030008E*. Tokyo: Graduate School of International Corporate Strategy (ICS).

——. 2006a. "Learning and the Self-Renewing, Network Organization: Toyota and Lexus Dealers." See pp. 27–40 in *Japan, Moving Toward a More Advanced Knowledge Economy. Volume 2: Advanced Knowledge-creating Companies*, edited by Takeuchi Hirotaka and Tsutomu Shibata. Washington, DC: The World Bank

——. 2006b. "Innovation Based on New Platform Strategy: Toyota's IMV (Innovative/International Multi-purpose Vehicle) Project." See pp. 1–39 in *Hitotsubashi University ICS Case Studies A060001E*. Tokyo: Graduate School of International Corporate Strategy (ICS).

Osono, Emi, Norihiko Shimizu, and Hirotaka Takeuchi. 2008. *Extreme Toyota: Radical Contradictions that Drive Success at the World's Best Manufacturer*. Hoboken, NJ: John Wiley & Sons.

Ozaki, Haruo. 2007. "China's Manufacturing Future Rests on the Development of Original Technologies." See pp. 1–3 in *JCER Staff Report* (September). Tokyo: Japan Center for Economic Research (JCER). Online available: <http://www.jcer.or.jp/eng/pdf/kenho070901e.pdf> (accessed October 21, 2008).

Ozawa, Terutomo. 1999. "Pacific Economic Integration and the 'Flying Geese' Paradigm." See pp. 55–91 in *Deepening Integration in the Pacific Economies: Corporate Alliances, Contestable Markets and Free Trade*, edited by Alan M. Rugman and Gavin Boyd. Northampton, MA: Edward Elgar.

Panasonic. 2007. "News & Events 10 January 2007 – Headquarters' News – Panasonic and Toray to Build the World's Largest Plasma TV Plant with a Capacity of One Million Units per Month." *News and Events* (January). Tokyo: Panasonic. Online available: <http://panasonic.co.jp/corp/news/official.data/data.dir/en070110-12/en070110-12.html> (accessed November 12, 2008).

——. 2008a. "About Panasonic." See pp. 1–2 in *Company Webpage*. Singapore: Panasonic, Singapore. Online available: <http://panasonic.com.sg/web/ccatId/CatPanasonic/1> (accessed November 12, 2008).

——. 2008b. "Panasonic R&D Center China." Beijing: Panasonic Corporation of China (PCC). Online available: <http://panasonic.cn/prdcc/english/research/index.jsp> (accessed November 12, 2008).

——. 2008c. "Matsushita Electric Industrial Co., Ltd. Annual Report 2008: We Are Panasonic." In *Panasonic Global*. Tokyo: Panasonic. Online available: <http://panasonic.net/ir/annual/2008/index.html> (accessed November 12, 2008).

Pempel, T. J. 2005. "Introduction: Emerging Webs of Regional Connectedness." See pp. 1–28 in *Remapping East Asia: The Construction of a Region*, edited by T. J. Pempel. Ithaca: Cornell University Press.

Perrucci, Robert. 1994. *Japanese Auto Transplants in the Heartland: Corporatism and Community*. New York: Aldine de Gruyter.

Pescosolido, Bernice A., and Beth A. Rubin. 2000. "The Web of Group Affiliations Revisited: Social Life, Postmodernism, and Sociology." *American Sociological Review*, Vol. 65 (February): 52–76.

Pilat, Dirk. 2006. "The Globalization of Value Chains and the Changing Nature of Manufacturing – Evidence, Implications and Policies." Tokyo: MITI, Research Institute of Economy, Trade and Industry (RIETI).

Piore, Michael J., and C. F. Sabel. 1984. *The Second Industrial Divide: Possibilities for Prosperity*. New York: Basic.

Polanyi, Michael. 1967. *The Tacit Dimension*. London: Routledge and Kegan Paul.

Porter, Michael. 1990. *The Competitive Advantage of Nations*. New York: Macmillan.

Portes, Alejandro. 1998. "Social Capital: Its Origins and Applications in Modern Sociology." *American Sociological Review* 24: 1–24.

Powell, Walter W. 1990. "Neither Market Nor Hierarchy: Network Forms of Organization." See pp. 295–336 in *Research in Organizational Behavior*, edited by L. L. Cummings and Barry M. Staw. London: Jai Press Inc.

Putnam, Robert D. 1993. *Making Democracy Work: Civic Traditions in Modern Italy*. Princeton, NJ: Princeton University Press.

Radosevic, Slavo. 1999. *International Technology Transfer and Catch-up in Economic Development*. Cheltenham, UK: Edward Elgar.

Ramstatetter, Eric. 2007. "Chapter 11: Thailand." *EAEP, East Asian Economic Perspectives* 18, 1 (February): 123–34.

Reed, Electronics Research. 2007. *Yearbook of World Electronics Data 2007, Volume 2: America, Japan & Asia Pacific*. Oxford: Reed Electronics Research.

Regnier, Philippe. 2004. "The East Asian Crisis and Beyond: New Perspectives for Linkages between SMEs and TNCs?" See pp. 160–9 in *Linking Local and Global Economies: The Ties That Bind*, edited by Carolo Pietrobelli and Arni Sverrisson. New York: Routledge.

——. 2005. "The East Asian Financial Crisis in Thailand: Distress and Resilience of Local SMEs." See pp. 148–62 in *Sustaining Growth and Performance in East Asia: The Role of Small and Medium Sized Enterprises*, edited by Charles Harvie and Boon-Chye Lee. Northampton, MA: Edward Elgar.

Renault. 2008. *Renault Atlas September 2008*. Paris: Renault.

Reuters. 2008. "Samsung and Sony Agreed on Establishing Additional 8th Generation Line at S-LCD." London: Thomson Reuters. Online available: <http://www.reuters.com/article/pressRelease/idUS80238+25-Apr-2008+RNS20080425> (accessed October 12, 2008).

Rhyu, Sang-young, and Seungjoo Lee. 2006. "Changing Dynamics in Korea–Japan Economic Relations." *Asian Survey* 46, 2: 195–214.

Rodrigo, G. Chris. 2001. *Technology, Economic Growth and Crises in East Asia*. Northampton, MA: Edward Elgar.

Sabel, Charles. 1989. "Flexible Specialization and the Re-emergence of Regional Economies." See pp. 17–70 in *Reversing Industrial Decline? Industrial Structure and Policy in Britain and Her Competitors*, edited by P. Hirst and J. Zeitlin. Oxford: Berg.

Sakakibara, Eisuke, and Sharon Yamakawa. 2004. "Trade and Foreign Direct Investment: A Role for Regionalism." See pp. 63–120 in *Global Change and East Asian Policy*

Initiatives, edited by Shahid Yusuf, M. Anjum Altaf and Kaoru Nabeshima. Washington, DC: World Bank Publications.

Sakamoto, Hiroshi. 2007. "Chapter 3: China." *EAEP, East Asian Economic Perspectives* 18, 1 (February): 27–50.

Samuels, Richard J. 1994. *"Rich Nation, Strong Army:" National Security and the Technological Transformation of Japan*. Ithaca: Cornell University Press.

Samuels, R. J., and E. Heginbotham. 1998. "Mercantile Realism and Japanese Foreign Policy." *International Security* 232: 171–203.

Saneh, Chamarik, and Susantha Goonatilake. 1994. *Technological Independence: The Asian Experience*. Tokyo: United Nations University Press.

Sasuga, Katsuhiro. 2004. *Microregionalism and Governance in East Asia*. London: Routledge.

Schaede, Ulrike. 2000. *Cooperative Capitalism: Self-regulation, Trade Associations, and the Antimonopoly Law in Japan*. New York: Oxford University Press.

——. 2003. "Industry Rules: From Deregulation to Self-Regulation." See pp. 191–214 in *Japan's Managed Globalization: Adapting to the 21st Century*, edited by Ulrike Schaede and William Grimes. Armonk, NY: M. E. Sharpe.

Schaede, Ulrike, and William Grimes. 2003. "Permeable Insulation and Japan's Managed Globalization." See pp. 243–58 in *Japan's Managed Globalization: Adapting to the Twenty-first Century*, edited by Ulrike Schaede and William Grimes. Armonk, NY: M. E. Sharpe.

Segal, Adam. 2005. "Chinese Techno-nationalism: Autonomy and Security in the World Economy." See pp. 205–22 in *Japan and China in the World Political Economy*, edited by Saadia M. Pekkanen and Kellee S. Tsai. New York: Routledge

Seth, Michael J. 2002. *Education Fever: Society, Politics, and the Pursuit of Schooling in South Korea*. Honolulu, HI: University of Hawaii Press and Center for Korean Studies, University of Hawaii.

SFIC. 2005. "Product Feature Determines Market Performance: Shanghai Panasonic Plasma Display Co. Ltd." In *Investment Shanghai: Foreign Enterprises*. Shanghai: Shanghai Foreign Investment Commission (SFIC), Shanghai Foreign Economic Relation and Trade Commission (SFERTC). Online available: <http://www.investment.gov.cn/2005-10-27/1130425544110.html> (accessed November 21, 2008).

Shimizutani, Satoshi, and Yasuyuki Todo. 2007. "What Determines Overseas R&D Activities? The Case of Japanese Multinational Firms." *RIETI Paper Series* 07-E-010. Tokyo: Research Institute of Economy, Trade and Industry (RIETI). Online available: <http://www.rieti.go.jp/en/publications/summary/07030009.html?mode=print> (accessed April 21, 2008).

Shiowattana, Prayoon. 1990. "Japanese Technology Transfer in Thailand." See pp. 110–31 in *Japan in Thailand*, edited by Kunio Yoshihara. Kyoto: Center for Southeast Asian Studies, Kyoto University.

Simon, Denis Fred. 1997. "Techno-security in an Age of Globalization." See pp. 3–21 in *Techno-security in an Age of Globalization: Perspectives from the Pacific Rim*, edited by Denis Fred Simon. Armonk, NY: M. E. Sharpe.

Simon, Denis Fred, and Jun Yongwook. 1997. "Technological Collaboration in the Semiconductor Industry: The Case of the Hitachi–Goldstar Strategic Alliance." See pp. 201–25 in *Techno-security in an Age of Globalization: Perspectives from the Pacific Rim*, edited by Denis Fred Simon. Armonk, NY: M. E. Sharpe.

Sklair, Leslie. 2001. *The Transnational Capitalist Class*. Oxford, UK: Blackwell.

Smitka, Michael. 1991. *Competitive Ties: Subcontracting in the Japanese Automotive Industry*. New York: Columbia University Press.

——. 1998. *The Textile Industry and the Rise of the Japanese Economy*. New York: Garland Publishing.

Sohn, Dong-won, and Martin Kenney. 2007. "Universities, Clusters, and Innovation Systems: The Case of Seoul, Korea." *World Development* 35, 6: 991–1004.

Solis, Mireya. 2005. "From Iron Doors to Paper Screens: The Japanese State and Multinational Investment." See pp. 108–29 in *Japan and China in the World Political Economy*, edited by Saadia M. Pekkanen and Kellee S. Tsai. New York: Routledge.

Sonoda, Mitsuhiro. 2007. "Mitsuhiro Sonoda, Head of Toyota Motor Thailand and the Thai–Japanese Chamber of Commerce, Discusses His Take on the Local Auto Industry." *The Nation – Business Page*, May 23, 2007. Bangkok: Nation Multimedia.

Ssangyong, Motor. 2008. "SsangYong Motor: History." Seoul: Ssangyong Motor. Online available: <http://www.smotor.com/en/overview/history.jsp> (accessed November 15, 2008).

STECO. 2006. "Interview with STECO President Park Jae-wook: Solidifying World-best LDI Package Supplier Position." *Journal of Korean Electronics* (April): 1. Online available: <http://www.gokea.org/neweiak/admin/upload/200604_1.pdf> (accessed November 12, 2008).

Steinfeld, Edward S. 2004. "China's Shallow Integration: Networked Production and the New Challenge for Late Industrialization." *World Development* 32, 11: 1971–87.

Steinmueller, W. Edward. 2002. "Knowledge-based Economies and Information and Communication Technologies." *International Journal of the Social Sciences* 171 (March): 141–54.

Storper, Michael. 1997. *The Regional World: Territorial Development in a Global Economy*. New York: Guilford Press.

——. 2000. "Globalization and Knowledge Flows: An Industrial Geographer's Perspective." See pp. 42–62 in *Regions, Globalization, and the Knowledge-based Economy*, edited by John H. Dunning. New York: Oxford University Press.

Sturgeon, T. J. 2002. "Modular Production Networks: A New American Model of Industrial Organization." *Industrial and Corporate Change* 11, 3: 451–96.

Sturgeon, Timothy J., and Richard K. Lester. 2003. "The New Global Supply-base: New Challenges for Local Suppliers in East Asia." In *MIT IPC Working Paper* 03–06. Cambridge, MA: Massachusetts Institute of Technology.

Sukonthapan, Pisawat. 2005. "Thai Intellectual Property Law." See pp. 219–47 in *Comparison of Industrial Agglomerations Between Asia and the Other Regions*, edited by Akifumi Kuchiki, Sun Jian and Juan J. Palacios. Chiba, Japan: Institute of Developing Economies (IDE), Japan External Trade Organization (JETRO).

SWAK. 2008. "Statistics on Spinning and Weaving 2008." Seoul: Spinners and Weavers Association of Korea (SWAK). Online available: <http://www.swak.org/eng/status.html?sub_top=top3> (accessed November 12, 2008).

Tachiki, Dennis S. 2005. "Between Foreign Direct Investment and Regionalism: The Role of Japanese Production Networks." See pp. 149–69 in *Remapping East Asia: The Construction of a Region*, edited by T. J. Pempel. Ithaca, NY: Cornell University Press.

TAI. 2005. "Thailand Automotive Institute Annual Report 2005." Bangkok: Thailand Automotive Institute. Online available: <http://www.thaiauto.or.th/About_Us/annual_eng_2005.pdf> (accessed October 12, 2006).

——. 2008. "Vehicle and Parts Exports (1996–2008)." Bangkok: Thailand Automotive Institute (TAI). Online available: <http://www.thaiauto.or.th/Records/eng/vehicleandpartexport_eng.asp> (accessed November 2, 2008).

TAIA. 2008. "Vehicle Production." Bangkok: Thai Automotive Industry Association (TAIA). Online available: <http://www.thaiauto.or.th/Records/eng/vehicleproduction_eng.asp> (accessed November 12, 2008)

Taniguchi, Makoto. 2007. "The Restructuring of Japan's East Asian Strategy." *ERINA Report* 75 (May): 8–10.

Taylor, Michael, and Simon Leonard. 2002. "Approaching 'Embeddedness.'" See pp. 1–18 in *Embedded Enterprise and Social Capital: International Perspectives*, edited by Michael Taylor and Simon Leonard. Burlington, VT: Ashgate

Tho, Tran Van. 1992. *Sangyō Hatten to Takokuseki Kigyō [Industrial Development and the Multinational Companies]*. Tokyo: Toyo Keizai Shimbunsha.

THTI, Thailand Textile Institute. 2007. "Thai Textile Statistics 2006/2007." Bangkok: Thailand Textile Institute (THTI). Online available: <http://www.thaitextile.org/th/download/stat06book.asp> (accessed November 4, 2008).

Tilly, Charles. 1998. "International Communities, Secure or Otherwise." See pp. 397–412 in *Security Communities*, edited by Emanuel Adler and Michael Barnett. Cambridge: Cambridge University Press.

TKS, Toyo Keizai Shimbunsha. 2007a. *Kaigai Kigyō Sōran Shinshutsu 2007: Kaishabetsuhen [Survey of Enterprises Investing Abroad 2007: Companies]*. Tokyo: Toyo Keizai Shimbunsha (TKS).

——. 2007b. Kaigai Shinshutsu Kigyō Sōran 2007: Kunibetsuhen [Survey of Enterprises Investing Abroad 2007: Countries]. Tokyo: Toyo Keizai Shimbunsha (TKS).

——. 2008. *Japan Company Handbook – Autumn 2008*. Tokyo: Tokyo Keizai Shimbunsha.

TMC, Toyota. 2008. "Toyota Annual Report 2008: Driving to Innovate New Value." In IR Library. Tokyo: Toyota Motor Company. Online available: <http://www.toyota.co.jp/en/ir/library/annual/pdf/2008/index.html> (accessed November 2, 2008).

TN. 1999. "Shiro Mizutani: Expert Suggests Specialized SMEs." *The Nation – Business*, July 5, 1999. Bangkok: The Nation Multimedia Publishing.

——. 2006a. "Matsushita Upbeat on Local Investment." *The Nation – Business*, October 4, 2006. Bangkok: Kwanchai Rungfapaisarn, The Nation Multimedia Publishing.

——. 2006b. "Thailand to Be R&D Center." *The Nation – Business*, December 27, 2006. Bangkok: The Nation Multimedia Publishing.

——. 2007a. "Pulling Double Duty – Mitsuhiro Sonoda, Head of Toyota Motor Thailand and the Thai–Japanese Chamber of Commerce." *The Nation – Business*, May 23, 2007. Bangkok: Nation Multimedia Publishing. Online available: <http://nationmultimedia.com/2007/05/23/business/business_30034917.php> (accessed May 23, 2007).

——. 2007b. "Much at State If Chinese Bubble Were to Burst: Thailand's Impressive Export Performance Is Intrinsically Linked to Sustainable Growth by Its Giant Northern Neighbor." *The Nation – Business*, May 28, 2007. Bangkok: Srisukkasem, Anoma, The Nation Multimedia Publishing. Online available: <http://nationmultimedia.com/2007/05/28/business/business_30035353.php> (accessed May 28, 2007).

——. 2007c. "Panasonic Cutting Costs as Exports Hit." *The Nation – Business*, June 14, 2007. Bangkok: The Nation Multimedia Publishing. Online available: <http://nation multimedia.com/2007/06/14/business/business_30036784.php> (accessed June 14, 2007).

——. 2007d. "Electrical/Electronics Sector Hit by Strong Baht." *The Nation – Business*, July 11, 2007 Bangkok: The Nation Multimedia Publishing. Online available: <http://nationmultimedia.com/2007/07/11/business/business_30040357.php> (accessed July 11, 2007).

——. 2007e. "Sony Focuses More on IT." *The Nation – Business*, July 16, 2007. Bangkok: The Nation Multimedia Publishing.

——. 2007f. "Second Milestone Auto Goal – Industry 'Can't Reach 3-Year Output Target.'" *The Nation – Business*, September 13, 2007. Bangkok: Chalida Ekvitthayavechnukul, Nation Multimedia Publishing. Online available: <http://nationmultimedia.com/2007/09/13/business/business_30048756.php> (accessed September 13, 2007).

——. 2008. *Japan Company Handbook: First Section Autumn, 2008*. Tokyo: Toyo Keizai Shimbunsha (TKS).

TNI, Thai-Nichi Institute of Technology (TNI). 2007. "Cooperation between METI and the Thai-Nichi Institute of Technology (TNI)." See pp. 1–8 in *METI News Release: 14 March 2007*, edited by METI. Tokyo: Japanese Ministry of Economy, Trade and Industry (METI). Online available: <http://www.tni.ac.th/download/080404TNI.pdf> (accessed November 4, 2008).

Todeva, Emanuela. 2006. *Business Networks: Strategy and Structure*. London: Routledge.

Todo, Yasuyuki, and Satoshi Shimizutani. 2007. "Overseas R&D Activities and Home Productivity Growth: Evidence from Japanese Firm-level Data." In *RIETI Paper Series* 07-E-008. Tokyo: Research Institute of Economy, Trade and Industry (RIETI). Online available: <http://www.rieti.go.jp/en/publications/summary/07030006.html?mode=print> (accessed April 4, 2008).

Toray. 2006. "Toray – Global Network – Press Release August 23: Toray to Establish a New Polypropylene Spunbond Company in China." Tokyo: Toray. Online available: <http://www.toray.com/news/fiber/nr060823.html> (accessed March 5, 2008).

——. 2008a. "Annual Report 2008: April 1, 2007 – March 31, 2008." In *Toray Investor Relations – Archives – Released Presentation and Reports: 9 May 2008*. Tokyo: Toray Industries. Online available: <http://www.toray.com/ir/pdf/lib/lib_a246.pdf> (accessed October 12, 2008).

——. 2008b. "Toray Saehan Inc. – Webpage – Office and Plant – Overseas – China." Seoul: Toray Saehan. Online available: <http://www.toraysaehan.com/eng/company/market/market02.aspx> (accessed November 4, 2008).

TSMA, Thai Synthetic Fiber Manufacturers' Association. 2008. "Thai Synthetic Fiber Manufacturers' Association Profile." Bangkok: Thai Synthetic Fiber Manufacturers' Association (TSMA). Online available: <http://www.thaitextile.org/tsma/profile.php> (accessed November 12, 2008).

TSST. 2008. "Webpage – About TSST – Company Overview." Suwon, Korea: Toshiba Samsung Storage Technology Corporation. Online available: <http://www.tsstorage.com/korea/en/company/overview.php> (accessed October 12, 2008).

Tsuneishi, Takao. 1990. "Nikkei Kigyō Shudō ni yoru Gijutsu Iten [Technology Transfer among Japanese Affiliated Firms]." See pp. 161–98 in *Ajia no Kōgyōka to Gijutsu Iten [Technology Transfer in Asia's Industrialization]*, edited by Takao Taniura. Tokyo: Ajia Keizai Kenkyūjo [Institute of Developing Economies].

Ueno, Hiroshi, Takashi Murakoso, and Takumi Hirai. 2006. "Supplier System and Innovation Policy in Japan." See pp. 137–51 in *Small Firms and Innovation Policy in Japan*, edited by Cornelia Storz. New York: Routledge.

UN. 2008. *Comtrade Database*. New York: United Nations Statistics Division. Online available: <http://comtrade.un.org/db/default.aspx> (accessed November 12, 2008).

UNCTAD. 2001. *World Investment Report 2001: Promoting Linkages*. New York: United Nations Publications, United Nations Conference on Trade and Development (UNCTAD).

——. 2006a. "World Investment Report 2006: FDI from Developing and Transition Economies – Implications for Development." New York: United Nations Conference on Trade and Development (UNCTAD).

——. 2006b. *World Investment Report 2006, Country Fact Sheets*. New York: United Nations Publications, United Nations Conference on Trade and Development (UNCTAD).

——. 2008a. *Major FDI Indicators: Series – Direct Investment in Reporting Economy (FDI Inward)*. New York: United Nations Publications, United Nations Conference on Trade and Development (UNCTAD). Online available: <http://stats.unctad.org/FDI/TableViewer/tableView.aspx> (accessed November 12, 2008).

——. 2008b. *Interactive Database – Sources and Definitions – FDI Flows*. New York: United Nations Publications, United Nations Conference on Trade and Development (UNCTAD). Online available: <http://www.unctad.org/Templates/Page.asp?intItemID=3201&lang=1> (accessed November 12, 2008).

——. 2008c. *Interactive Database – Sources and Definitions – FDI Stock*. New York: United Nations Publications, United Nations Conference on Trade and Development (UNCTAD). Online available: <http://www.unctad.org/Templates/Page.asp?intItemID=3202&lang=1> (accessed November 12, 2008).

Unger, Danny. 1998. *Building Social Capital in Thailand: Fibers, Finance, and Infrastructure*. New York: Cambridge University Press.

UNIDO. 2008. *International Yearbook of Industrial Statistics 2008*. Vienna: United Nations Industrial Development Organization (UNIDO).

Urata, Shujiro. 2006. "The Changing Patterns of International Trade in East Asia." *Creation of New Contemporary Asian Studies, Working Paper #30* (October). Tokyo: Waseda University.

Urata, Shujiro, and Kazutomo Abe. 2005. "The Impact of Japan–Korea FTA and CU on Economy, Industry and Trade in Japan and Korea." See pp. 158–200 in *Korea–Japan FTA: Toward a Model Case for East Asian Economic Integration*. Seoul: Korea Institute for International Economic Policy.

Uzzi, Brian. 1996. "The Sources and Consequences of Embeddedness for the Economic Performance of Organizations: The Network Effect." *American Sociological Review* 61 (August): 699–718.

Wang, Hongying. 2001. *Weak State, Strong Networks: The Institutional Dynamics of Foreign Direct Investment in China*. Oxford: Oxford University Press.

Wang, Jenn-huan. 2006. "China's Dualist Model on Technological Catching-up: A Comparative Perspective." *The Pacific Review* 19, 3 (March): 385–403.

Ward's, Automotive Group. 2007. *Ward's Automotive Yearbook 2007*. Southfield, MI: Ward's Automotive Group.

——. 2008a. *Ward's Automotive Yearbook 2008*. Southfield, MI: Ward's Automotive Group.

——. 2008b. *Motor Vehicle Facts & Figures 2008: Documenting the Performance and Impact of the US Auto Industry*. Southfield, MI: Ward's Automotive Group, a division of Penton Media Inc.

Weiqiao. 2008. "Weiqiao Textile Announces 2008 Interim Results." *The Earthtimes Newspaper*. Hong Kong: The Earthtimes Co. Online available: <http://www.earthtimes.org/artiles/show/weoqoap-textile-announces-2008-interim-results.541947.shtml> (accessed November 2, 2008).

Whitley, Richard. 1996. "Business Systems and Global Commodity Chains: Competing or Complementary Forms of Economic Organization?" *Competition and Change* 1: 411–23.

Whittaker, D. Hugh, and Robert E. Cole. 2006. "Conclusions and Reflections: Emergent Models." See pp. 298–320 in *Recovering from Success: Innovation and Technology Management in Japan*, edited by D. Hugh Whittaker and Robert E. Cole. New York: Oxford University Press.

Wijnhoven, Fons. 2006. "Knowledge Management: More Than a Buzzword." See pp. 1–16 in *Knowledge Integration: The Practice of Knowledge Management in Small and Medium Enterprises*, edited by Antonie Jetter. New York: Physica-Verlag.

Willet, Susan. 2001. "Introduction: Globalization and Insecurity." See pp. 1–12 in *Special Issue of IDS Bulletin – The Value of Value Chains: Spreading the Gains from Globalization*, edited by Gary Gereffi and Raphael Kaplinsky. Sussex, UK: Institute of Developing Societies.

WIPO. 2006a. *Statistics on Patents – Patents Granted by Office, 1883–2005 (Breakdown by Resident and Non-Resident)*. Geneva: World Intellectual Property Organization (WIPO). Online available: <http://www.wipo.int/ipstats/en/statistics/patents/> (accessed October 15, 2006).

——. 2006b. "Exceptional Growth from North East Asia in Record Year for International Patent Filings." In *WIPO Press Release 436 (2/23/2006)*. Geneva: World Intellectual Property Organization (WIPO). Available online: <http://www.wipo.int/edocs/prdocs/en/2006/wipo_pr_2006_436.html> (accessed September 15, 2007).

——. 2007. "Information by Country – China." See pp. 1–2 in *Country Profile from WIPO Guide to Intellectual Property Worldwide*, edited by World Intellectual Property Organization WIPO. Geneva: World Intellectual Property Organization (WIPO). Online available: <http://www.wipo.int/export/sites/www/about-ip/en/ipworldwide/pdf/cn.pdf> (accessed May 5, 2007).

Wong, Kar-yiu, *et al.* 2004. *Northeast Asian Economic Integration: An Analysis of the Trade Relations among China, Japan, and South Korea*. Seoul: Institute for Monetary and Economic Research, Bank of Korea.

Woolcock, Michael. 1998. "Social Capital and Economic Development: Toward a Theoretical Synthesis and Policy Framework." *Theory and Society* 27: 151–208.

World Bank. 2006a. "Table 5.10 The Information Age." See pp. 302–6 in *2006 World Development Indicators*, edited by World Bank. Washington, DC: The World Bank.

——. 2006b. "Science and Technology." See pp. 306–9 in *World Development Indicators 2006*, edited by World Bank. Washington, DC: World Bank.

——. 2008. *World Development Indicators Database, April 2008*. Washington, DC: World Bank.

WTO. 2007. *International Trade Statistics*, Merchandise Trade by Product, 2006: Table II.64 Leading Exporters and Importers of Textiles. Geneva: World Trade Organization (WTO). Online available: <http://www.wto.org/english/res_e/statis_e/its2008_e/its08_merch_trade_product_e.htm> (accessed April 15, 2007).

Xinhua. 2007. "Toyota Motor Corporation Chairman Appointed Japan–China Economic Association Head." *People's Daily Online*, March 23, 2007. Beijing: Xinhua News Agency. Online available: <http://English.people.com.cn/200703/23/print20070323_360193.html> (accessed April 5, 2008).

Yamashita, Shoichi. 1995. "Japan's Role as a Regional Technological Integrator and the Black Box Phenomenon in the Process of Technology Transfer." See pp. 338–56 in *The Emerging Technological Trajectory of the Pacific Rim*, edited by Denis Fred Simon. Armonk, NY: M. E. Sharpe.

Yamazawa, Ippei. 1988. "Increasing Imports and Structural Adjustment of the Japanese Textile Industry." *The Developing Economies* Vol. 18, No. 4 (December): 441–62.

——. 1990. *Economic Development and International Trade: The Japanese Model*. Honolulu, HI: East–West Center, Resource Systems Institute.

——. 2000. "Summary Report: Towards Closer Japanese–Korean Economic Relations in the 21st Century." Tokyo: Research Planning Department, Institute of Developing Economies (IDE). Online available: <http://www.ide.go.jp/English/Lecture/Sympo/pdf/kankoku_soron.pdf> (accessed October 12, 2006).

Yokota, Kazuhiko. 2007. "Chapter 2: Japan." *EAEP, East Asian Economic Perspectives* 18, 1 (February): 13–26.

Yoshimatsu, Hidetaka. 2002a. "Japan and Industrial Adjustment in Asia: Overproduction Problems in the Synthetic Fiber Industry." *Pacific Affairs* 75, 3 (Fall): 377–98.

——. 2002b. "Preferences, Interests, and Regional Integration: The Development of the ASEAN Industrial Cooperation Agreement." *Review of International Political Economy* 9, 1 (March): 123–49.

——. 2003. *Japan and East Asia in Transition: Trade Policy, Crisis and Evolution, and Regionalism*. New York: Palgrave Macmillan.

——. 2006. "Industrial Competition and Technology Standards: Techno-regionalism in East Asia." In *RCAPS Working Paper* No. 06–1 (June). Beppo, Japan: Ritsumeikan Center for Asia Pacific Studies (RCAPS).

Yoshizaki, Tatsuhiko. 2007. "Japan, the Quiet Genius: Toward a Strategic Public Diplomacy." Gaiko Forum: Japanese Perspectives on *Foreign Affairs* 7, 1 (Spring): 21–7.

Zhang, Yun-ling. 2007. "The Feasibility of Establishing an East Asian FTA: A Chinese Perspective." In *Occasional Paper Series* 07–5. Seoul: Institute for Global Economy.

Index